with particular reference to fisheries | a balanced science of renewable resources

henry a. regier

a washington sea grant publication
distributed by university of washington press
seattle and london

The series of lectures on which this publication
is based were supported by grant number 04-5-158-48
from the National Oceanic and Atmospheric
Administration to the Washington Sea Grant Program.

The U.S. Government is authorized to produce and
distribute reprints for governmental purposes
notwithstanding any copyright notation that
may appear hereon.

Library of Congress Cataloging in Publication Data

Regier, Henry A., 1930-
    A balanced science of renewable resources.

    (A Washington sea grant publication ; WSG 78-1)
    "Based on ten lectures presented in April 1976
at the University of Washington, College of
Fisheries."
    Bibliography: p.
    1.   Fishery management--Methodology.
2. Conservation of natural resources--Methodology.
3. Renewable natural resources--Management.
4. Ecological research--Methodology. I. Title.
II.   Series.
SH328.R43   333.9'5   78-4979
ISBN 0-295-95602-X

preface

These ten essays are based on ten lectures presented in April 1976 at the University of Washington, College of Fisheries. Much of the material appeared first in papers published elsewhere, mostly in collaboration with a number of colleagues, and listed in the references at the end of this series. What is new here is the attempt to integrate the concepts and show how they compare with a variety of recent writings by others.

Besides those listed in the references, many other colleagues have helped in the development of these concepts and arguments. Often their assistance occurred indirectly in the course of some group activity. Rather than name scores of individuals, I have acknowledged those organizations that were particularly helpful for me. The institutions and agencies for which I have worked directly are the Ontario Ministry of Natural Resources, Cornell University, University of Toronto, Food and Agriculture Organization of the United Nations, Canada Fisheries and Marine Service, and the Fisheries Research Board of Canada. In addition I have benefited particularly by involvement with the Institute of Ecology, FAO's Advisory Committee on Marine Resources Research, UNESCO's and Canada's Man and the Biosphere Programs, the University of Washington's College of Fisheries, the University of Wisconsin's Laboratory of Limnology at Madison, the U.S. Council of Environmental Quality, the American Fisheries Society, and Statistics Canada. My thanks to the many friends who have helped. If I have miscast any of their contributions, I would be pleased to hear of it.

This set of essays contains many quotes. Special permission was requested for the more extensive quotes, and was granted as follows: Section 2.3, Prof. Ian I. Mitroff; Section 2.4, Prof. Magoroh Maruyama; Sections 4.2 to 4.4, Dr. G. Römer of E. Schweitzerbart'sche Verlagsbuchhandlung; Sections 5.2, 5.5 and 10.4, Prof. Erich Jantsch; Sections 7.4 and 7.5, Dr. S.P. Bakker of W. Junk b.v. - Publishers; Sections 6.2, 8.3, 8.4, 8.5 and Chapter 9, Dr. J.C. Stevenson of the Journal of the Fisheries Research Board of Canada. These and other authors and editors are not responsible for the way in which I have used their works.

I am grateful to Mrs. Vera Howland for clerical, secretarial, and editorial assistance far beyond the call of duty.

If this work were in a more polished and respectable state, I would dedicate it to my wife Lynn and our daughters Chris, Andrea, and Susan. They have missed out on many family outings, weekends, and holidays because I was too much involved in these matters.

H.A.R.

foreword

In 1971 a grant was given by the Washington Sea Grant Program to support innovative teaching at the College of Fisheries, University of Washington. One way was the initiation of a lecture series on resource management to be given by distinguished scientists in this field, both national and international.

Dr. Regier follows such eminent speakers as Dr. J.Gulland, Dr. D. Cushing, Dr. W.E. Ricker, and Dr. G. Hempel. While the earlier series dealt with specific topics in population dynamics, the present lecture series enlarges the scope by examining the concepts of renewable resource management as they have developed during the last three decades. The single species approach has been replaced by an ecosystem approach, and the static deterministic stock model by a process-oriented one which together provide an understanding of the dynamics of the whole ecosystem. However, resource management is not based solely on biological criteria because utilization decisions have sociological and economic implications. In essence, resource management must be seen as part of a political process.

Dr. Regier's own career and research reflect some of the changes and attitudes in resource management. He earned his doctorate from Cornell University in 1961 and began his university teaching there. Later, he transferred to the University of Toronto, where he now is professor of zoology and also associate director of the Institute of Environmental Studies.

Dr. Regier's voluminous production began in 1960 and dealt at first with problems in small pond management and the statistics of sampling and estimation. These themes were soon enlarged to include the dynamics of populations in large lakes, and lately to the ecology of fishing and other stresses on aquatic communities. Now, new facets have been added. The encroachment of civilization on the environment calls for some form of control on industrialization and urbanization together with a fisheries rehabilitation policy. Dr. Regier belongs to a small but growing group of scientists who are trying to mobilize the innovative forces of society to meet the present challenges of the use of natural aquatic resources which include the accountability of resource managers, the solution of conflicts arising from multiple resource user groups and the acceptance of rights of low consumption exploitation. All this calls for new inquiry systems and new modes of public discussion and participation in decision making.

Over the years, the international community has repeatedly called on Dr. Regier for many contributions. He was chief of the Stock Evaluation Branch of the Food and Agricultural Organization of the United Nations from 1970 to 1971. There he participated in international training courses on marine pollution, served as chairman of a working party on ecological indices, represented FAO at the first meeting of Global Investigations of Pollution in the Marine Environment, and was a member of the Advisory Committee on Marine Resources and Research, to mention a few assignments. As a keynote speaker at international meetings, he is known to examine topics and ideas far beyond the framework in which fishery biologists usually operate.

The present volume represents a contribution and an early attempt to formalize in writing ideas and thoughts which began fermenting in North America and elsewhere in the Sixties. The concepts presented in these ten lectures in all probability will be looked on as truisms during the next decade.

Ole A. Mathisen
University of Washington
February 1978

contents

1

when solutions become problems

1.1 Introduction

People who are expert on such matters tell us that western science derives from classical Greek logic which was in large part universalistic, homogeneous, hierarchic, reductionist, linear, quantitative, and competitive (Maruyama 1974; Jantsch 1975). It has been dominated by a one-way causal paradigm. Within this dominant stream there were a number of currents and cross-currents. Mitroff and Pondy (1974) have sketched the characteristics of five of these and have attached to each the name or names of some leading exponents of each approach: Leibniz, Locke, Hegel, Kant, and Churchman-Singer (see Chapter 2). The latter pair may in fact be sufficiently different from the main stream of western science as to be compatible with another stream.

Some of the philosophers of science suggest that our current difficulties in coming to grips with environmental matters, or coming to terms with nature (ours included), can be traced back to the presuppositions of classical Greek logic. They suggest that we might consider instead the logical systems of the Navajo, Japanese, and Black Africans and seek to develop a science based on their logics, or on as much as is shared by those three cultures. Here the emphasis might focus on heterogenous more intuitive, non-hierarchic networks and on holistic, non-linear systemic and qualitative concepts (Maruyama 1974). In short, these people hold to a "mutual causal paradigm".

These two paradigms do not complete the set (see Chapter 2), but I consider this pair to be sufficient for immediate purposes. It seems to me that a viable short-term option--say for the next two decades--is to try to develop a flexible, balanced approach by conserving what is useful in western science and by nurturing one or more of its radical complements. An alternative may eventually achieve a position of practical utility approaching that of conventional science.

The yin-yang model of the East may be used to summarize this introduction (Figure 1.1). Here yang can be identified very approximately with western science, and yin with the radical alternative sketched above (Japanese - Navajo - Black

Figure 1.1 Yin-yang model.

African). Certain approaches to systems studies, as in naturalistic ecology, may be perceived as part of the yin element within the predominant yang of western science.

In this book I intend to work at the fringes of western science, where ecology is to be found, and hope to show that we have the conceptual raw material for a more flexible and balanced science of renewable resources than is commonly applied in practice.

Which sets the question: Can social institutions and decision makers accommodate forms of science other than that based on classical Greek logic? According to the experts, the short answer to that is yes. In fact they are gradually becoming more accommodating (see Chapter 10).

1.2 Difficulties with Extensions of Conventional Approaches

What with the U.S. National Environment Policy Act (NEPA) of 1969, the Stockholm Environment Conference (1972), and the subsequent founding of the United Nations Environmental Program (UNEP), and many other important events, ecologists should be elated and confident. Instead many are

frustrated, querulous, and unsure of themselves. Why?

Let us examine four major practical initiatives of the last ten years:

-- NEPA's environmental impact assessment process

-- various regional and national attempts at environmental monitoring

-- regional planning, particularly as related to urbanization

-- International Commission for Northwest Atlantic Fisheries approach to total allowable catches and to negotiated quotas.

All of these initiatives are widely perceived as being imperfect, perhaps to the extent of being impractical.

For some ecologists, the remedy is to approach all of them through dynamic simulation that rests predominantly on a quantitative understanding of some of the major proximate one-way causal mechanisms involved. Though necessary, even the best one-way approach will be found to be insufficient. Planning and decision making is very much an iterative, step-wise mutually-influencing process; the information particularly relevant to various stages and various groups may be at very different scales of resolution and of different spatial, temporal, and ecological scope. To depend on one dynamic simulation to satisfy all or most of these needs is simply to opt for prohibitively expensive ecological input.

Of course, everyone recognizes this, but why don't we act on it? Why don't we seek to develop effective iterative stepwise collaborative approaches to

2

ecological decision-making that relate well to the more comprehensive decision processes of society? Why do we settle for one-shot inputs when we know that such an approach is inefficient? Let us examine these initiatives in more detail.

1.3 NEPA and Environmental Impact Statements

Problems with U.S. National Environmental Policy Act (NEPA) of 1969 include the following:

a) "The law's instructions for preparing an impact report are not specific enough to ensure that an agency will fully or even usefully, examine the environmental effects of the projects it plans." (Gillette 1971).

b)" ... there is presently no uniformity in approach or agreement upon objectives in an impact analysis..." (Leopold et al. 1971).

c) "...the current project-by-project approach to environmental impact assessment does not allow for considerations of cumulative impact of multiple projects or actions in a given geographic area." (Jenkins et al. 1974).

d) "Court decisions are as likely to be determined on technicalities as on the merits of cases." (Jenkins et al. 1974).

At a June 1975 seminar (held in Ann Arbor and sponsored by Argonne National Laboratory and the University of Michigan) on

"biological significance" with respect to impact assessment I sensed that some ecologists were beginning to fear that the environmental assessment process had become a nightmare. The process was consuming financial and ecological manpower resources out of all proportion to the amount of useful insight generated. (Of course the measure of success need not include contribution of ecological insight: it might be sufficient that the process ensure that the environment be less degraded than it would have become in the absence of the impact assessment process.)

A quick look at the presuppositions and design of the Leopold matrix (Leopold et al. 1971) may illustrate some of the beginnings of the nightmare.

a) The environmental impact assessment report was to be fed into the planning process rather late (see Figure 1 of Leopold et al. op. cit.): by implication this delay assigns environmental and renewable concerns to a rank of secondary importance.

b) Coming late in the planning process, perhaps the only role left to the ecologist is that of an adversary, particularly if "the preparer (of the impact assessment statement) will consider all impacts to be potentially deleterious because all the (beneficial) factors would have been covered in the engineering report (that preceded the environmental impact statement)" (Leopold et al. op. cit.)

c) The one environmental impact statement must address issues at

3

various levels of scope and scale: whether the overall project should proceed at all; if so, which of the major alternatives is least deleterious; how can design changes make the least deleterious process even less damaging. To expect solutions to such problems is asking too much of one short-term impact assessment.

d) It turns out that the Leopold matrix is perhaps best suited to assist with fine-tuning of a primary decision made on the basis of other non-ecological considerations. In settling for this approach the ecology establishment may in effect have gained relatively little.

Yorque (1975), a report on a workshop in Vancouver, listed eight "common myths about environmental impact assessment," EIA, as follows:

- EIAs should consider <u>all</u> possible impacts of the proposed development.

- Every new impact assessment is unique and must be designed as though there were no relevant background of principles, information, comparable past cases.

- Comprehensive state of the system surveys (check lists, etc.) are a necessary first step in EIA.

- Detailed descriptive studies within subsystems can be integrated by systems analysis to provide overall understanding and predictions of systems responses (impacts).

- Any good scientific study is useful for decision making.

- Physical boundaries based on watershed units or political jurisdictions can provide sensible limits for impact investigations.

- Systems analysis will allow effective selection of the best alternative from several proposed plans and programs.

- Development programs can be viewed as a fixed set of actions (e.g., a one-shot investment plan) which will not involve extensive modification, revision, or additional investment as program goals change over time and unexpected impacts arise.

Apparently we should not read these so-called myths as necessarily in error. But taken singly they can be seriously misleading.

Most of these misgivings are not new, Jenkins et al. (1974) advised as follows:

"Recommendation 5. Revise agency guidelines for environmental impact assessment with CEQ guidelines (Federal Register, Aug. 2, 1973) to require that consideration of and research on environmental effects of projects commence at the outset in planning technological developments, resource utilization, environmental modification, construction of human settlements, and other human activities."

"Recommendation 6. (a) Maximum practicable use of quantitative analysis, prediction, and evaluation of ecological impacts;

(b) Post-impact evaluation of predicted impacts and unpredicted consequences on a selected basis; and

(c) Identification and initiation of research of problems amenable to solution by additional ecological research."

These recommendations were carried somewhat further at the SCOPE Workshop on Impact Studies in the Environment (Munn 1975).

To digress briefly: these recommendations are a challenge to ecologists. We are rapidly developing the capability to meet that challenge. At least we could quickly achieve it if the various ecological and practical innovators were more open to the suggestion that they could each benefit from more collaborative interaction. By analogy with the jargon relating to the logistic curve as applied to ecosystem succession, ecology may soon be through its "r phase" and will then enter its "K phase". The "r phase" is dominated by a lot of competitive opportunists, the "K phase" by more organized symbionts.

1.4 Environmental Monitoring and Indices

Perhaps the most widely known and most controversial approaches to environmental accounting involve the use of single aggregated indices incorporating data from a wide variety of variables (see Curlin 1973, Dee et al. 1973, Inhaber 1974). Their major strength is also their major weakness—a single measure is produced. The elements to be aggregated can each be measured with precision satisfactory to the natural scientist concerned about bias and accuracy. When aggregated, an index may connote little more than there is or is not cause for immediate concern depending on the day.

A Planning Committee on Environmental Indices of the U.S. National Academy of Sciences (NAS) (Eisenbud et al. 1974), reported as follows:

A rather extensive nationwide network of air pollution monitoring sites exists. Instruments and methods have been developed to measure routinely a number of air pollutants including carbon monoxide, oxides of nitrogen, oxidants, sulfur dioxide and particulate matter. However, only sulfur dioxide and particulate matter are routinely monitored in most places, the number of monitoring sites is often inadequate, and they are not always located to provide representative information. . . .

A number of indexing schemes have been devised and promulgated during the past several years. These have been devised either locally for local use or under a federally-sponsored contract for more general use. Because most of them were developed locally, there now exists a multitude of air quality indices—some good and some bad—but none so outstanding as to gain wide acceptability. A systematic appraisal of the available air quality indices is needed for the purpose of introducing uniformity.

Most air quality indices, whether they be local or national in scope, rely on the two classes of federal ambient air quality standards. The primary standards are those which are designed to protect the public health, whereas the more stringent secondary standards are intended to protect materials, vegetation, and the amenities. These standards have been used to provide a basis for aggregating the major air pollutants for indexing purposes, but there are many air pollutants for which there are as yet no federal standards.

Aggregation of air quality measures into an index is generally accomplished by dividing the measured ambient air pollutant levels by the federal standard for each pollutant (scaling), weighting these ratios according to importance, and then arithmetically or geometrically averaging the ratios. A major practical difficulty here is that the time intervals specified in the federal standards are not uniform and do not in fact agree with the time intervals usually used to collect data on pollutant concentrations. A more severe problem is that the relative importance or weights given to the several pollutants are assigned arbitrarily in most instances.

The same NAS committee also addressed water quality:

Construction of water indices is similar conceptually to that of air indices but is more difficult in application for two main reasons.

First, the number of known pollutants in water exceeds that in air, and not all are expected to be found in all water bodies. Thus, the indexing method becomes quite involved, even without consideration of synergisms.

Second, no single use exists for water for which we can set standards, as we can for air because of our prime concern with its effect on human health. The federal Environmental Protection Agency considers water to have five separate, but not necessarily mutually exclusive, uses for which quality criteria can be established: recreation and aesthetics; public water supplies; fish, other aquatic life, and wildlife; agricultural use; and industrial use. The 1972 Water Quality Act Amendments set a goal of water quality sufficient to allow for water-contact recreation and the propagation of fish and shellfish. Thus, there is no single criterion to form the basis of a water quality index.

In spite of the difficulties, several water indices are available . . .

The aggregated index approach seems to appeal particularly to conventional engineers with primary interests in *construction.* For them the ecological environment is often viewed as an adversary to be overcome. They may acknowledge that the holistic natural environment has properties of positive importance, but are unwilling to address that prospect in any detail. For their purposes some aggregated index of environmental quality is deemed to be sufficient. After all the information will usually be used to set constraints on the engineers' activities--better from their viewpoint to have one generalized constraint than a series of locally-calibrated specific constraints.

Many ecologists are unhappy with such procrustean approaches to the environment, but their resistance is weak, and seeking viable alternatives is not a priority research issue with them.

1.5 Regional Planning with Overlays

Geographers, architects, as well as urban and regional planners, lean heavily on spatial maps for information summary, analysis, and transfer (Abler et al. 1971; McHarg 1969; Burton et al. 1974).

Separate maps, perhaps in the form of transparent overlays, deal with different variables that are presumably each of some implicit or direct interest to the clients. These variables may each be normalized, as with some of the agglomerated indices referred to above, and symbols or numbers may be used to summarize data on and over the map. If in addition the overall interest can be addressed as a simple linear function of the separate variables, then a single map can be produced--thus a spatially-oriented analogue of the temporally-oriented condensed index for monitoring. The strengths and weaknesses of such condensed homogenized information in the form of maps are similar to those of time series of agglomerated indices. The approach appeals particularly to developers and their allies in bureaucracies, perhaps for reasons similar to those specified for engineers in the preceding section.

Table 1.1 compares the "overlay" approach to impact assessment dealing both with maps of single variables as well as of aggregated indices, with the matrix approach and with an aggregated index (Battelle) approach used heretofore mostly for monitoring.

1.6 International Commission for Northwest Atlantic Fisheries Total Allowable Catches

The International Commission for Northwest Atlantic Fisheries (ICNAF), is perhaps the most advanced among the numerous international fisheries commissions of the world. The 1975 approach to management involved scientific estimation of total allowable catches (TACs) on a stock-by-stock basis

with subsequent political negotiation as to national quotas. Regier and McCracken (1975) summarized events in the preceding decade as follows:

Development of management regulations in the ICNAF has been described in some detail and with various kinds of emphases by Möcklinghoff, Edelman and Dokuchaev, Bogdanov and Konstantinov, and Hennemuth (all in Ricker and Weeks 1973). In general their accounts suggest that relatively early in this period it was recognized that — with the rapid growth of the fishery, changes in techniques, and development of "mixed" fisheries — minimum mesh size in itself was insufficient to regulate the fishery. In fact, because much fishing was being carried on for small fish forms previously "unexploited" (silver hake, red hake, herring, etc.) minimum mesh size regulation could at best be only partially effective.

Scientists became concerned about the rapid development of fishing pressure before administrators did, for example, in the haddock and herring fisheries. Scope for action by managers was constrained by the differing aims of the various countries and by the ICNAF convention which, though it allowed the setting of overall catches, did not provide for allocation by countries. Without the possibility of allocating by countries, effective regulation was not feasible. Haddock became a special case because the stock had been so reduced that it was of interest only to Canada and the United States.

With changes in the ICNAF protocol (1969) to allow countries to negotiate quota allocations, controls have been introduced very rapidly. Estimates of total allowable catches (TACs), determined by scientists on a stock by stock basis, have been made available to

Table 1.1 COMPARISON OF THREE GENERAL APPROACHES.

		Leopold[1]	Overlay[2]	Batelle[3]
Capability	Identification	Medium	Medium	High
	Prediction	Low	Low	High
	Interpretation	Low	Low-medium	High
	Communication	Low	High	Low-medium
	Inspection procedures	Low	Medium	Low-medium
Action complexity capability		Incremental alternatives	Fundamental and incremental alternatives	Incremental alternatives
Risk assessment capability		Nil	Nil	Nil
Capability for flagging extremes		Low	Low	Medium
Replicability of results		Low	Low-medium	High
Level of detail	Screening of alternatives	Incremental	Fundamental and incremental	Incremental
	Detailed assessment	Yes	Yes	Yes
	Documentation stage	Yes	Yes	Yes
Resource requirements	Money	Low	Maps low; computer high	High
	Time	Low	Maps low; computer high	High
	Skilled manpower	Medium	High	High
	Computational	Low	Maps low; computer high	Medium
	Knowledge	Medium	Medium	Medium

Source: Munn, 1975.

[1] Leopold matrix especially for ecological impact assessment.

[2] Transparent map overlays, especially for regional planning.

[3] Agglomerated condensed indices, especially for monitoring.

ICNAF negotiators. To illustrate the rapid development of this approach, in 1968 there were no stocks under TAC control, yet by 1974 TACs had been set for some 56 stocks in the ICNAF region. Most stocks in the Canadian region are now under TAC control (excepting such species as dogfish, skate, sand launce), even some of those in deep water. TACs have been set with a number of aims in mind, such as rehabilitation of haddock or other stocks, preventing takes beyond the maximum sustained yield for various species, preventing excessive pulse fishing, and taking precautionary measures for developing fisheries such as capelin.

TACs for individual stocks must include the "incidental" species catch. To solve this problem by indirect methods, as well as limit total effort, a TAC for all species combined was set at a lower level than the sum of its parts.

In addition to TAC and quota control, ICNAF countries introduced an enforcement scheme that allowed mutual inspection of catches and gear at sea. The scheme was far from perfect, but did provide a basis for ensuring that regulations were being followed.

Recapitulating then, the situation prior to 1976 in the ICNAF region was: Total landings maintained by addition of new species such as capelin, silver hake, grenadiers; greater exploitation of more traditional species such as redfish, flatfish, mackerel, herring; greater fishing intensity on the major cod stock off the east coast of Newfoundland; and attempts to control the fishery using a TAC quota approach on a stock by stock basis.

In a broader context the Advisory Committee on Marine Resources Research (ACMRR) of the U.N. Food and Agri-culture Organization (FAO) advised in part as follows in 1975 under the heading "Some actual problems in the management of advanced fisheries:"

The Committee took note of the increased problems which were occurring as fishing intensified and was exerted on a growing number of the components of the ecosystem. The problems of optimizing yields from species complexes and/or reducing the incidental take of unwanted species was considered.

The Committee considered the management and assessment problems associated with various kinds of multi-species fishery situations occurring in different parts of the world. It recognized that multispecies fisheries occur in their most extreme form in tropical areas where the exploited resources comprise large numbers of species, mostly small and having similar biological characteristics, living in close association, and caught together throughout each fishing area. In such fisheries, owing to the large number of species and their close association, management action on an individual species basis is mostly impracticable. Instead, for most of them their scientific assessment and management must be approached on a total fish biomass basis (or perhaps the biomass of the most important species) using data on total fish catch and fishing effort, and an appropriate assessment model. If the time frame is right, this procedure may take account of the complex biological interactions between the species within the multi-species population. Provided catch data for individual species, or groups of species, are available, adjustments can be made to allow for differences in their monetary value should the management objectives be in economic terms.

Because of the close association in such multi-species populations management measures aimed at controlling age of fish capture (e.g., mesh size) are for the most part impracticable. Management of such fisheries must therefore be effected by measures controlling the exploitation rate of the total fish biomass, either by catch quotas or by the direct regulation of fishing effort (including restrictions on the kinds and sizes of fishing gears and fishing methods).

Although less complex in terms of the number of species exploited by each unit fishery than those in the tropics, major multi-species fisheries, posing basically similar management and associated assessment problems take place in temperate fishing regions, notably in the North Atlantic and North Pacific. In these regions the post-war expansion of fishing activities has resulted not only in more intensive exploitation of the stocks of certain species of long standing economic importance but also in the development of new fisheries on species not previously fished, many of which are important components of the prey of the traditionally fished species.

We therefore need to broaden our understanding of the dynamics of whole ecosystems, and not merely of selected species. Appropriate methods of doing this, to permit construction of predictive models, must now be developed. The inadequacy of present understanding of ecological processes has important consequences also for management. Uncertainties in predictions of responses of the resources to changes in exploitation and to management measures must be taken into account in formulating and applying those measures. The existence of significant interactions, and of natural trends — as well as fluctuations — in resources productivity, and the fact that the ecosystems are not necessarily, even approximately, in "steady-states", means that the concepts of optimum or maximum sustainable yields must be applied with caution. This calls for more intensive and more varied research than has usually been required in the past.

When I articulated similar concerns during the period of 1967-1970 in seminars in Toronto, Ann Arbor, Seattle, Rome and elsewhere as well as in writing (Regier et al. 1969, Regier 1972) the recognized experts generally rejected them as misdirected, overblown or untimely. Research priorities were directed toward expanding and improving the details of the conventional approach. Even when ICNAF took new management initiatives that effectively rendered much of the conventional thinking obsolete, relatively little research was directed to examine alternatives already available. Marine fisheries workers are now in a scramble to provide appropriate scientific information for the new management realities.

1.7 Alternatives

I have presented evidence that some of the primary scientific and management methods still being employed in impact assessment, environmental monitoring, regional planning, and fisheries resource management are far from perfect. Furthermore, it might be very difficult to achieve major improvements simply by developing these further along conventional lines.

Various alternatives are already competing for attention. The overall process by which science and its application progresses is not entirely a chance process. In what follows here, some properties of that process are examined in the attempt to discover what may lie ahead.

10

2
presuppositions for a science of renewable resources

Western societies have developed some promising planning and decision mechanisms and processes, in the attempt to deal creatively with widely perceived difficulties stemming from the usual man-nature interaction. Judging from the evidence sketched in Chapter 1, science and technology seem to be lagging behind the implicit demand. Evidence is presented that a broader conceptual paradigm is emerging. If this happens, the conventional science may not be rendered obsolete, rather it may be taken up and used within a more comprehensive process.

In searching for a more useful basis, one may iterate between concepts too large for regional or local environmental issues and those too small. Scope and scale are important though difficult relevance criteria (Paulik 1972). Also one may assess the conventional offerings of various disciplines such as physical geography or oceanography, ecology, operational management, and political science. Furthermore, every science has a variety of alternative methodologies to offer.

2.1 Science and Technology under Attack

Science and technology have contributed greatly to the shaping of the kind of world in which we now live - for better and for worse. The last few decades have seen a tremendous expansion in research and development activity in all industrialized countries and, despite the fact that the main justifications in providing the necessary resources have been the objectives of defence, national prestige and economic growth, the expansion period has been one of somewhat unquestioning euphoria for science. This has now come to an end and both legislators and the public at large are questioning not only the costs and benefits of research but the very objectives which have induced its expansion (King 1973).

Perhaps the most revolutionary event of the last third of this century is the claim for social responsibility of science and technology. Up to now, technologies have been hailed and accepted as irresistible forces bringing unquestioned technical advantages and intrinsically containing seeds of economic and social progress.

Now, at the beginning of the 1970's, for the first time, new technological developments are being rejected - allegedly on social grounds.

From now on, the problem of technology will be raised in terms of benefits and dis-benefits, in other words, of social accounting of technological change. (Hetman 1973).

Whether or not scientists and technologists individually prefer to deal with broader social and moral issues, we will apparently have to justify ourselves more explicitly and be more accountable henceforth.

On balance this may be good news for most scientists interested in renewable resources and the natural environment. Many of our most difficult practical problems stem from the indirect and/or long-term effects of programs initiated as a result of the findings of other scientists applied in the absence of informed consideration of ecological and social consequences. But many ecologically-literate, practically-oriented scientists are now themselves chafing under the growing political impact from naturalists with tender consciences, ranging from preservationists sympathetic to various birds, turtles and sea mammals to humane pet-oriented people who seek to mitigate the misery of all animals.

2.2   Social Reorientation

Compared to years just past, the mid-1970's are turning out to be quite placid, at least in North America and Europe. Remember just a few years ago - unrest in large cities, antiwar sentiment, university sit-ins and riots, environmental activism, Ralph Naderism, women's lib, zero population growth, and anti-nuclear neo-Luddites.

Where have all the radicals gone? Far, far away? Some have opted for communes in the desert. Others have decided to help effect social change. Still others concur with Pogo: We have met the enemy and he is us.

Now it is common for people in authority to recognize that far-reaching changes are in process. In 1974, Maurice Strong, then Director-General of the U.N. Environment Program, issued a set of 10 prescriptions to "cure our ailing world" (Strong 1974). North American governments have not rushed to diagnose those ills cited by Strong, nor to prescribe comprehensive cures, but most politicians will nod sympathetically when approached on the subject. Meanwhile senior advisory and policy groups have taken up the call. Thus Jenkins et al. (1974) began their report on the role of ecology in the U.S. federal government:

As we approach our Nation's third century, we are facing the [1] rethinking and [2] redirection of national goals. There is recognition of the [3] finite nature of resource availability and of the [4] environment's capability to accommodate continued unlimited growth. There is also increasing awareness of our [5] reliance on a healthy environment, and growing recognition of the [6] profound impact of man's activity on all components of the environment.

I have added the numbers in square brackets to focus attention on three pairs of important aspects of this highly condensed introductory statement. These six aspects seem far less controversial now than they were 10 years ago. But the lack of intense controversy may be the result of broad realization that no one of high authority has yet clearly proposed any truly major changes in national policies anywhere in the Western world. The Canadian prime minister is widely renowned for his "musings" on the subject, but policy proposals are another matter.

Are we now in a comparative lull before an imminent storm? Or are we in the eye of the hurricane? The turmoil of the late 1960's may return with renewed vigor, but from an apparently different direction. We would be fortunate indeed if a major re-orientation of Western national goals could be achieved in relative calm. However, it would not be prudent to presuppose a tranquil passage, much as we might prefer that it turn out that way.

So the first criterion for a balanced science of natural resources is that it be robust and flexible enough to work under difficult social and political conditions. This need may relate more, in the first instance, to the field of operations design and management. It has important bearing on how research and management of renewable resources are undertaken. Fragile concepts, methods, and personalities should perhaps be stored away carefully in case the passage turns out to be quite rough.

2.3 Inquiry Systems

Mitroff and Pondy (1974) have shown how policy analysis and organization theory can be classified into a number of "inquiry systems". Perhaps the science related more directly to renewable resources, a level below formal policy analyses and organization theory, can also be addressed in part from this perspective.

The brief sketches of five inquiry systems (see below), selected by Mitroff and Pondy

from a larger set, may be addressed with the aid of a number of questions.

1. To *which* one of these is my own personal scientific approach most closely related?
2. Is my personal approach better *suited* to my broader scientific objectives than all the other approaches?
3. Do the workers with whom I *collaborate* share a similar approach?
4. If not, what kind of problems arise as a result of the *mismatch?*

With those questions as possible leads into the matter, here are five sketches:

(1) Leibnizian Inquiry Systems (ISs) "are the epitome of formal, deductive reasoning systems. No matter what the problem, Leibnizian ISs always strive to reduce it to a completely pure mathematical, logical, or formal representation of some sort." Unless this can be done, advocates of Leibnizian inquiry do not feel the 'true essence' of the problem situation is fully comprehended.

(2) Lockean ISs. "If Leibnizian inquiry represents a pure, abstract theory-before-data view of the world, then Lockean inquiry represents the exact reverse. Lockean ISs are the epitome of inductive, experiential systems. No matter what the problem, Lockean ISs always strive to break it down into its 'basic experiential components' on which one can make some 'direct observations.' Unless one can do this, those who advocate Lockean inquiry do not feel we have adequately understood or captured the "true essence" of the problem situation."

(3) Kantian ISs "emphasize a view of the world which gives equal emphasis to both (the Leibnizian and Lockean perspectives).

Table 2.1 PRACTICAL ASPECTS OF FIVE APPROACHES TO INQUIRY, HERE RELATED
IN PARTICULAR TO TECHNOLOGICAL FORECASTING AND ASSESSMENT.

Philosophical approach	Characteristcs of problem	Truth content	Guarantor notions
Leibniz			
theoretical, mathematical	- well defined - analytical - possible solution seen - deductive	- formal symbolic systems separate and prior to data component - true by definition	- specifies what shall count in systems - consistency - completeness - comprehensive - model builder believes in its accuracy
Locke			
empirical, realistic	- well-defined - experimental - consensus opinion possible - inductive	- experiential truth = agreement between observers on data - data prior to and separate from theory	- no prior presumption of theory (*tabula rasa*) - emphasis on real world where agreement can operate effectively
Kant			
synthetic, idealistic	- definable - defined objective - mixed analytic and experimental - alternative solutions sought - conceptualizing goals not included	- synthetic - every theoretical term must have empirical referent - must show how data relate to model - theory and data inseparable	- degree of FIT between underlying theory and data - aspirations differ, therefore many alternatives sought
Hegel			
dialectic, conflictual	- ill-defined - opposing objectives - intuitive or synthetic reasoning - on any issues, by definition, opposing sides	- truth conflictual - plan and counter-plan will elicit synthesis - without these, data are only partial - same data can support either theory	- opposing views help decision-makers to creative synthesis - contributes to redefining problem
Singer			
pragmatic, feasible	- ill-defined - unclear objective - multi-disciplinary - reflective reasoning	- truth--relative to goals and objectives - all 'laws' are only well-confirmed hypotheses - only legitimated by how *seriously* we take them - brings in ethics behind acceptance of laws	- all above systems included - all present data interact with normative thinking

Table 2.1 *(continued)* PRACTICAL ASPECTS OF FIVE APPROACHES TO INQUIRY, HERE RELATED IN PARTICULAR TO TECHNOLOGICAL FORECASTING AND ASSESSMENT.

Example	Forecasting	Underlying assumption	Weakness
- laws of physics - computer simulation models - OR--e.g., model of a transportation system - Forrester Meadows 'world model'	- correlation, analysis - substitution analysis (e.g., natural to synthetic fibers)	- processes can be described in terms of a model - models represent some aspects of reality; therefore can be fitted to model	- no way to justify assumptions - changing reality requires change in model
- statistics - economic models - scientific experiments	- trend extrapolation - consensus Delphis (early) - regression analysis - specific short-term developments	- data collection and data analysis separate and not related to formal theory - in Delphi--explicit degree of consensus measured	- not clear about criteria for "relevance" - choosing data implies theory
- philosophical inquiry - later Delphis (e.g., question of leisure in future)	- normative forecasting - gaming - cost/benefit analysis - scenarios - morphological analyses - cross impact matrices - relevance trees, decision trees	- no one 'best' solution - connection between theory and data - alternative solution may approximate truth	- not for problems with single-clear formulation - no guide for choice between solutions
- strategic planning - policy Delphi	- not much used here	- conflict will expose assumptions underlying opposing theories - can keep opposing arguments clear of data	- no assurance that synthesis is 'best' solution - conflict may not be resolved
- hard to find - almost a meta-system	- finding forecasting methodology that applies to a particular problem	- interrelatedness of systems	- too comprehensive

Source: Mitroff and Turoff, 1973, as adapted by Janice Tait. Used with permission.

Kantian ISs emphasize that it is impossible to collect data on a phenomenon without having to presuppose some theory, no matter how implicit that theory may be, with respect to the phenomenon being investigated."

(4) Hegelian ISs "are designed to actively seek the strongest possible conflicting views of a problem. It is hoped that, as a direct result of witnessing the conflict and the discrepancy between opposing interpretations of the same data, the decision maker will be in a better position to observe for himself the crucial role assumptions play in the analysis of problems. Hegelian or Dialectical ISs are founded on the presumption that conflict is a better principle than agreement in exposing the hidden assumptions on which every representation of the world depends."

(5) Singerian-Churchmanian "ISs are complex functions of all the previous ISs, and there are an infinite number of ways of combining these ISs. Singerian-Churchmanian ISs are the epitome of integrative interdisciplinary systems. (They) are anti-reductionist (and) do not believe that the process of inquiry can be reduced to a single set of fundamental entities on which knowledge can be shown to depend. From (their point of view), each of the previous ISs is nowhere nearly as independent or as pure as their proponents would have us believe . . . A whole systems view of the world implies that in the strict sense there are no such things as "components" which have an existence independent of other components and of the whole system itself. Under this philosophy, it is wrong to perceive any single component as more critical than any other, and hence, it is wrong to allocate all of one's inquiry efforts to studying one phase of the system over any other. . . . Singerian-

Churchmanian inquiry arises directly out of the tradition of American pragmatism. For the pragmatist, the "truth" of a model is not to be identified solely with its formal structure or with formal truth tests of the model, but rather with the ability of the model to effect significant social action, i.e. implementation."

The papers by Mitroff, Turoff, and Pondy include much more detail that relates these various inquiry systems more closely to environmental and resource issues, though more at the policy level than at the level of a project or local problem. Thus they point out that some dynamic simulation modelers, at least at the policy level, tend to be Leibnizian in perspective; earlier versions of Delphi procedures were Lockean; recent versions of Delphi methods are Kantian; some aspects of legal processes and some environmental activists are Hegelian; and American political pragmatists may be Singerian-Churchmanian.

More specifically addressing fisheries science, the various schools of population dynamics, whether biological (maximum sustainable yield) or economic (optimum sustainable yield) tend to be Leibnizian in outlook. Fish taxonomists and ethologists tend to be Lockean. Interactions between groups of national scientists in international commissions may often be Hegelian--or chaotic. Interdisciplinary groups of advisors to Western governments, or working under FAO auspices, may seek to work in a Kantian mode. But ultimately it seems that the big national decisions, even on international issues, are taken pragmatically, in a Singerian-Churchmanian way.

Would the Lockean or Leibnizian biologists and economists approach their work differently if they understood more about the different logics that predominate in the various places in which their information and insights are to be used? if so, *should* they be influenced, or is it their role simply to gather wood and haul water, i.e., to supply unprocessed scientific raw material? Should they be participants, or mere informers? My own predisposition is to seek participation; others may well settle for being instruments in decision-making if this will permit them time to address the questions that are of primary interest to them. Clearly we need specialists who know circumscribed subject matter in depth. We also need others who have a fairly clear insight into how the information will be used.

2.4   Major Cultural Alternatives

Early in the present century the work of Max Weber, Albert Schweitzer and others started western intellectuals on a search for the deep presuppositions and the dynamic essence of their civilization. The origins of current ethics, religions, science and technology, and their dynamic interactions, were of prime interest. Varieties of western approaches were analyzed comparatively. As the overall Western outlook came to be understood, it was compared with major alternatives, such as the Eastern or Animistic. Now, some seventy years later, many thoughtful people recognize some relevance in these studies. Even scientists, having survived a half-century of positivistic anti-intellectualism, are now perusing this kind of literature.

The summary by Maruyama (1974) shown in Table 2.2 presents three options in a highly condensed form. In North America, it would be difficult if not impossible to find strict adherents to a single option; nevertheless, there are socially important groups that in some general way hold predominantly to one of the set. In communicating with others a scientist adhering to the more conventional unidirectional causal paradigm may encounter great difficulty, especially if he fails to perceive the depths of the conceptual differences that may occur.

It is because of such differences that the inquiry systems related to the important political questions cannot be limited to those that seem adequate for physical scientists (see Section 2.3).

2.5   A Glance at Fisheries Science

In the classical scientific tradition, a statistical test, if appropriately designed and executed, can show that a false hypothesis is indeed false. Classically, not just any hypothesis that might be presented is worth testing--it should be derived deductively from a theory that purports to be "true" and has the explicit or tacit support of some respected people. If a logically deduced hypothesis can be shown to be false, then the parent theory cannot be true as it stands. On the other hand, that theory which survives a growing number of critical tests gains in favor. This approach is consistent with the Leibnizian inquiry system sketched in section 2.3.

It is widely held that new theory is usually the product of individual "creativity", i.e. the process of successful theorizing is not well understood. Once the theory exists, hypotheses may be deduced and tested in the classical manner. It may take a long time to devise a good test--some biologists say that Darwin's theory of evolution hasn't yet been well tested.

All this as preamble to a series of questions: What are the theories of, say, conventional fisheries science? If they exist, have they been tested? Which, if any, have survived numerous tests?

Table 2.2 CHARACTERISTICS OF THREE PARADIGMS ALL OF POSSIBLE INTEREST TO SCIENTISTS STUDYING OR MANAGING RENEWABLE RESOURCES AND THE NATURAL ENVIRONMENT.

	Unidirectional causal paradigm	Random process paradigm	Mutual causal paradigm
Science	Traditional "cause and effect" model	Thermodynamics; Shannon's information theory	Post-Shannon information theory
Information	Past and future inferrable form	Information decays and gets lost; blueprint must contain more information than finished product.	Information can be generated. Nonredundant complexity can be generated without preestablished blueprint
Cosmology	Predetermined universe	Decaying universe	Self-generating and self-organizing universe
Social organization	Hierarchical	Individualistic	Non-hierarchical interactionist
Social policy	Homogenistic	Decentralization	Heterogenistic coordination
Ideology	Authoritarian	Anarchistic	Cooperative
Philosophy	Universalism	Nominalism	Network
Ethics	Competitive	Isolationist	Symbiotic
Esthetics	Unity by similarity and repetition	Haphazard	Harmony of diversity
Religion	Monotheism	Freedom of religion	Polytheistic harmonism
Decision process	Dictatorship, majority rule, or consensus	Do your own thing	Elimination of hardship on any single individual
Logic	Deductive, axiomatic	Inductive, empirical	Complementary
Perception	Categorical	Atomistic	Contextual
Knowledge	Believe in one truth. If people are informed they will agree	Why bother to learn beyond one's own interest?	Polyocular: must learn different views and take them into consideration.
Methodology	Classificational, taxonomic	Statistical	Relational, contextual analysis, network analysis
Research hypothesis and research strategy	Dissimilar results must have been caused by dissimilar conditions. Differences must be traced to conditions producing them.	There is probability distribution; find out probability distribution.	Dissimilar results may come from similar conditions due to mutually amplifying network. Network analysis instead of tracing of the difference back to initial conditions in such cases.
Assessment	"Impact" analysis	What does it do to *me?*	Look for feedback loops for self-cancellation or self-reinforcement
Analysis	Pre-set categories used for all situations	Limited categories for his own use	Changeable categories depending on situation
Community people viewed as	Ignorant, poorly informed, lacking expertise, limited in scope	Egocentric	Most direct source of information, articulate in their own view, essential in determining relevance
Planning	By "experts;" either keep community people uninformed, or inform them in such a way that they will agree	Laissez-faire	Generated by community people

Source: Maruyama, 1974.

Now for some rhetorical questions about models addressed to a particular set of interrelated events, at a particular site, in a non-replicated period. If some a priori theory exists that purportedly applies to the "event" (in the aggregated, unitary sense), it should be possible to deduce a site-specific hypothetical model for the particular issue of interest, and then test it for realism using relevant data. But what can be done in the absence of a relevant theory? In the classical view, one can simply guess a theory, with a very low probability that it will survive testing or if it survives that it can be generalized in time and space-- if based only on the experience of the one event. Is a theory that may only apply to one site during one specific period really worthy of the name "theory"?

It seems to me that a classical theory-based approach to resource science is possible, but demands a strategy different from that used by most fisheries workers. For example, theory relevant to *fisheries systems as such* can be intuited, perhaps most readily by comparing and contrasting relatively gross events and large scale processes in a number of separate resource systems that appear to bear some broad resemblance to each other. The "compare and contrast approach" may be as old as science (cf. long-term observation of one process or active experimentation) but has been sadly neglected by most fisheries workers.

A related approach to theory formulation involves a search for analogues. From an appropriate perspective, a fisheries resource system may bear some similarity to another system for which theory has already been developed. So, if they exist, the theories that apply to systems of lakes, or reservoirs, or rivers may perhaps be "extended" to other aquatic systems

for which no theory has yet been developed.

I labor these matters because proponents of "hard science" in the Leibnizian mode, particularly physical scientists, often decry the "soft" methods of pragmatic decision makers. Yet when they address large scale issues they either do so inadequately by studying isolated components or ignore the classical well-tested methods of hard science in favor of less rigorous methods. It seems to me that Leibnizians cannot have it both ways.

2.6 Holism and Reductionism

Those who work within a holistic context presuppose that the whole is more than its parts (if it can be perceived to have parts at all) and that the whole is ultimately more "real" than the parts. The reductionists disagree and assume instead that ultimately reality is approached by reducing an aggregate into its smallest "real" entities.

There are epistemologies other than reductionism and holism, but these two are of recurring interest to western scientists. One can perhaps guess a reason for the interest by noting that western scientists presuppose a hierarchic universe. In rejecting monotheism as ultimate reality at the top, also rejecting the possibility that *Homo sapiens* is the prime mover, the third most likely possibility may have been to seek reality in the irreducible elements of the soil, air, water, and fire. Whatever the reason may have been, the basic hierarchic perception isn't the only one possible, as Maruyama (1974) has pointed out (see above).

A lot of rather tiresome posturing occurs between workers at adjacent levels of the perceived hierarchy. Those who work higher up believe that they are more "powerful", while those lower down fancy that they are closer to the "truth". In fisheries too.

"Holism" in the sense of Maruyama's mutual causal paradigm is more interesting than the connotation of "holistic" with which I might describe the scientist addressing a higher level than I am addressing in the hierarchy of a western world view. It is an intriguing possibility that American pragmatism, as developed in the Singerian-Churchmanian inquiry system by Mitroff and Pondy (1974, see above), might be able to accommodate without great difficulty arguments derived from within a mutual causal paradigm. This is something that should be of particular interest to scientists studying renewable resources and the natural environment.

Incidentally some traditions in medicine seem to be closer in spirit to a mutual causal paradigm than to a one-way causal paradigm. We might search there for theoretical analogues to renewable resource systems.

2.7   A Manifesto

Society's demands for scientific information and insights will be very different in the 1980's from what they were in the 1960's. The various scientific disciplines now most firmly ensconced are singly and jointly poorly organized to face the challenges of the 1980's.

On matters of the ecological environment and renewable resources, North American societies are now creating and developing innovative planning and decision-making processes. Environmental assessment review, regional land-use planning, urban planning, strategic planning, direct public consultation, intergovernmental "reference groups", are examples. The scientific insights and information currently made available to workers in these new institutional roles are often irrelevant, usually poorly integrated, and seldom readily assimilable. Minor tinkering with "communication skills" will not suffice.

The above statements are not in any sense radical or new. Innovations have found expression in the creation of "environmental" schools or faculties, institutes at universities, major private institutes (International Institute of Applied Systems Analysis), international programs (Man and the Biosphere of UNESCO), and private consulting firms. Most of these initiatives have encountered major difficulties and the risk of collapse is ever present.

Such innovations are often characterized as being "interdisciplinary". Unless such attempts are quite explicitly derived from one or more holistic, transcendent perspectives (from the viewpoint of the more conventional disciplines), they are unlikely to rise above an ad hoc and ineffectual multidisciplinarity. To be effective, interdisciplinarity must pass the test that the contributions of different workers cannot readily be recognized as characteristic of the separate disciplines (see Chapter 6).

It follows that we need to achieve clarification of at least one overriding principle *or* some sufficient set of more partial but mutually congruent concepts. Such a unitary or aggregated perspective would then provide a basis for a breakthrough into better science for future work related to the ecological environment and renewable resources. If one broad concept does become clarified it is likely that dissidents will seek to develop and clarify an alternative, which may lead to a productive dialectical interchange.

An "aggregated" option is taken here and the following four perceptions are proposed as a sufficient set to get the process started.

(a) Nature is a *system* with some measure of self-regulatory and evolutionary capability that can be crippled to its eventual detriment. Appropriate methods of systems analysis may be applied to expose this system's properties. An evolutionary systems perspective (compare von Bertalanffy (1962) with Jantsch (1975) or Maruyama (1974)) may be found here and there among workers in all the conventional scientific disciplines but is not sufficiently well developed in any one of them. But this kind of *commonality* needs to be exposed and assimilated into the new approach that is aimed specifically at our concerns. Thus a new paradigm may be a transdisciplinary synthesis with emergent properties and not a completely radical alternative.

(b) While the approach of reductionist decomposition has enabled man to design and construct highly complex systems, this same approach is of limited value when applied to the study of man's effect on ecological systems. Ecosystems and social systems are both strongly coupled and are thus essentially nondecomposable; yet, decomposability is a fundamental western technological assumption. Thus man has the technological tools that enable him to produce even greater environmental changes; yet these same tools are of limited value in predicting the ecological significance of such changes. Man is faced with an environmental predicament which can be stated as follows: Man's ability to modify the environment will increase faster than his ability to foresee the effects of his activities. This predicament rests not on western man's lack of interest in his environment (though this is a contributing factor) but on his relative inability to understand complex organized ecosystems, i.e., the environmental predicament rests ultimately on man's capacity, rather than on his will (Bella and Overton 1972).

(c) Within the newly emerging modes of planning and decision making (see above), the innovators are specifying problems and challenges pragmatically in *general language* that does not prejudge the relevance of particular disciplines. Once it seemed that unless an issue were stated in the quantitative idiom of a certain school of economics, or as an engineering challenge it could be dismissed as unformulated. The planning and decision-making structures and processes themselves are now more nearly transdisciplinary in essence and practice than has been the case heretofore.

(d) A practical role for ethics is being rediscovered. Stewardship, conservation, equity, reverence for life and other concepts are attracting renewed interest and commitment. These qualities may be congruent with a scientific *systems* perspective, perhaps within a mutual causal paradigm, and can be made relevant within the emerging political institutions.

People with this kind of viewpoint (four points above, but often not formulated explicitly) can be found scattered all through society and its various institutions and organizations. It may now be the right time to consolidate these perspectives and competences into fully operational capabilities.

3
life history of an ecological paradigm

3.1 Why "Paradigm"?

Mature or world-weary workers may show impatience upon encountering the word "paradigm" or other new-fangled jargon. It might be dismissed as just another neologism--a new term for an old concept. In circles where the term is novel some folk may be misled into thinking that new insight is involved; thus warn the skeptics. Why is "paradigm" needed when such concepts as broad perspective, conceptual framework, perceptional model, philosophical principle, observational focus or scientific idiom might do equally well?

Trivially, paradigm is one word while the others each involve two--hence, some gain in conciseness.

Not so trivially, the word paradigm has gained wide currency recently because of work by Kuhn (1970) on the nature of scientific revolutions. He dealt with the dynamics of science as influenced *inter alia* by social processes operating within the community of scientists. In this respect, science as a process is not very different from philosophy, religion, or politics--much as some scientists would prefer that it be otherwise.

If I judge correctly most scientists who use "paradigm" nowadays do so in the context of Kuhn's paradigm.

Good-natured cynics may enjoy F.E.J. Fry's jibe of some years ago. He knew the worth of paradigms--they came at five for a dollar. Nevertheless he was subsequently celebrated in a *Festschrift* subtitled "The Fry Paradigm" (Lawrie and Kerr 1976).

3.2 An Example of Paradigm Evolution

The population dynamics approach has become "conventional wisdom" and has developed a paradigm of its own. Table 3.1 shows the temporal sequences of three series that were interrelated and contributed to the evolution of "fish population dynamics". Other events, particularly the emerging needs for scientific information and concepts in management and regulatory agencies, also contributed and are taken up briefly below.

In fisheries the term "population dynamics" has developed a fairly distinctive set of connotations.

Table 3.1 EVOLUTION OF POPULATION DYNAMICS IN FISHERIES SCIENCE

	Scientific events	Quantitative capabilities in fisheries science	University events in North America
PRE 1880	Taxonomy, anatomy, morphology and "natural history."		Biology departments introduced widely
1880	Relative abundance by CPUE, size at sexual maturity ("minimum size").		
1890	Aging by size classes, tagging, aging by scales and hard parts.	Abacus; slide rule; logarithms math tables; calculus.	Aquatic science specialists at Cornell University
1900	Species' life history, general effects of pollution.		U. of Wisconsin U. of Toronto
1910	Feeding habits--stereotyped or not? population responses to fishing up, size and age changes		U. of Michigan U. of Indiana, etc.
1920	Gear selectivity related to size, physiology, growth--determinate or indeterminate?		U. of Washington, College of Fisheries
1930	Migration--homing, population estimation--simple versions, physical factors and habitat niche, genetics, Gaussian statistics.	Gaussian statistics, mechanical adding machines, books of tables; mechanical back-calculator for growth estimation	Fish & Game Depts. in various agriculture colleges and universities
1940	Mortality rates--simple versions, population and mortality parameters--discrete probability distributions.	Discrete probability statistics using tables or calculation by hand; optical "scale readers."	Ecology institutionalized
1950	Stock-recruitment relationships--2 types, recruitment as function of habitat variables, deterministic simulation of self-regulatory population, MSY.	"Rapid" mechanical motor-driven calculators; acoustical and electronic sensing devices; small electronic computers; simple simulation; differential equations.	
1960	Interspecific population interactions, dynamic simulations of self-regulatory populations, total allowable catches in multi-spp. system.	Acoustical fish counting; bigger computers; computer software; automated data systems; well-trained computer slaves; complex simulations; specialized computer languages.	F.W.S. Co-op Units in U.S.A.
1970	Ecosystem simulations, e.g. IBP, dynamic simulation of management options.	Remote sensing; computer networks; remote terminals; pocket calculators	Sea Grants in U.S.A.

Internally it has integrated conceptual structure and methodological process, and externally it has recognized its distinctiveness from other scientific approaches. It may now be considered a proto-discipline and some protagonists might welcome a concerted movement to achieve full discipline status and recognition of its pre-eminent relevance to fisheries problems.

In the field of ecology generally, study of any aspects of process or flow at the population level of organization is termed "population dynamics" by some. In fisheries science the connotations of the term are narrower with some attempt made to keep them narrow.

The emergence of fish population dynamics as now perceived can be dated to about 1950. From the scientific viewpoint, the first simulations of sequences of complete life cycles that purported to be sufficiently realistic for practical purposes occurred about then. These were accomplished more or less concurrently at Lowestoft, Seattle, Toronto and perhaps elsewhere. Just prior to this time, workers of various disciplines joined forces quite effectively on fisheries matters, particularly biologists and statisticians.

From the practical viewpoint, a series of fisheries problems came to a head in the late 1940's. In the North Sea fisheries, the recovery of the fish stocks during the 1939-1945 war was evidence that stocks had been significantly overfished and would again be so affected--perhaps to the point of being seriously overfished. Some groundfish stocks of the Northwest Atlantic were recognized as intensively fished. The study of halibut of the Northeast Pacific had involved controversy--the Thompson-Burkenroad debate (Skud 1975). Salmon runs along northernmost coasts of North America were deteriorating. In continental North America the backbone of all fisheries management--the fish hatchery program-- had collapsed under the weight of scientific criticism. Such was the multi-faceted practical reality at the time of the emergence of the population dynamics paradigm about 1950.

3.3 Notes on Other Paradigms Related to Fisheries

The following sketches are examples, from a bigger set in ecology (see Chapter 4). Here the emphasis is on examples of fisheries-related approaches.

3.3.1 Linnean-Darwinian Approach

Sample personalities include--besides K. von Linne and C.R. Darwin--D.S. Jordan, C. Hubbs, G. Svärdson, E. Mayr, and many more. Almost all fisheries workers simply accept as a foregone conclusion that their investigations begin with classical taxonomic identification of species. Recently the need to do so always has been queried (Balon 1975).

3.3.2 Physiological Autecology or Physical Factors Ecology

Sample personalities include J. von Liebig, F.F. Blackman, V.E. Shelford, F.E.J. Fry (1971), P. Doudoroff, J.R. Brett, D.F. Alderdice, and many working on pollution bioassays and population. Empiric and theoretical works have developed in phase. (Lawrie and Kerr 1976).

3.3.3 Fish Yield in Lakes and Reservoirs

Sample personalities include D.S. Rawson, J.B. Moyle, K.D. Carlander, T.G. Northcote and P.A. Larkin, R.A. Ryder, and R.M. Jenkins. Within sets of lakes subject to relatively homogeneous climatic, physiographic, social and market forces, the average fish yield can be related to a series of readily measured whole-system variables. Theoretical work is now underway.

3.3.4 Balance in Ponds

Sample personalities include H.S. Swingle, R.O. Anderson, and Y.A. Tang. On the basis of experimental and empirical work various states of "balance" amongst fish species present were related to pond owner preferences and simple corrective measures were developed to remedy observed deviations from what was

desired.   Theoretical studies are now
underway.

3.3.5   Fish Taxocenes in Rivers

Sample personalities include M. Huet, R.
Cuinat,S.G. Kryzhanovsky, and E.K. Balon.
Different kinds of fish associations
(taxocenes) have been related to different
sets of abiotic features characteristic of
the various reaches of river, from
headwaters to mouth.   At the level of
river taxocenes as taxocenes, relatively
little theoretical work has been initiated.

3.3.6   Shellfish Communities

Sample personalities include C.G.J.
Petersen, V.A. Kostitzin, J. Hylleberg, W.
Stephenson, and V.F. Gallucci. Questions
of how the spatial patterns of shellfish
and other invertebrates inhabiting soft
tidal flats and estuaries are determined
have been explored in a number of
systems.

3.4   Assimilation of a Paradigm
into Practice

It is inevitably a complex process by
which a new scientific theoretical concept
and/or methodological approach becomes
clarified, is tested, gains acceptance by
peers, and is transmitted directly and by
education to practice, eventually to be
institutionalized as "conventional
wisdom".   Table 3.2 is a simplified,
idealized sketch of the process that may
be just realistic enough to be useful for
expository purposes

I have incorporated the usual classifica-
tions: (a) innovative science, transfer via
(b) education and training into (c) widespread practice, followed by social
acceptance to the extent that the idea as
elaborated becomes (d_1) standard practice
and the (d_2) practitioners become the
recognized authorities in the relevant
subject area.

The arrows depict the main stream of
innovative effort--they are not intended
to imply that Stage A is completed before
Stage B is initiated, etc.   A helpful
analogy might depict innovation in small
drops falling all over the space of Table
3.2, but initially in more concentrated
form in Stage A. A flood results with the
crest sweeping in the direction of the
arrows to Stage D.

The elements within each of the stages
are publication media types characteristic
of the different stages.   Those here
included are particularly relevant to
sciences related to renewable resources
and the natural environment.

Consider the current state of fish population dynamics with respect to Table 3.2.
Practical assimilation has apparently
reached into Stage D. Thus J.A. Gulland's
recent books (e.g. Gulland 1972) present
standard models and methods; K.D.
Carlander's (1969, 1977) handbooks are
compendia of parameter values for North
American fish species; FAO's species
synopses perform a somewhat similar
function with broader geographic scope;
W.E. Ricker's (1975) third edition of
methods is widely accepted as a standard
collection.   There is no professional
association specifically bounded by
population dynamics concerns though
these appear to be dominant in the
American Institute for Fisheries Research
Biologists.   There has been interest in
establishing a specialized journal but this
has not happened, perhaps because
population dynamics papers have long
been accorded pride of place in journals
such as the Journal of the Fisheries

Table 3.2   SKETCH OF STEPS IN THE PRACTICAL ASSIMILATION OF A MAJOR
SCIENTIFIC PERSPECTIVE (PARADIGM), SHOWING THE TYPES OF
PUBLICATIONS CHARACTERISTIC OF DIFFERENT STAGES.

STAGE A:
CREATIVE INNOVATION   ←————————→   STAGE D:
INSTITUTIONALIZATION ————→

- Journal papers

- Bibliographies

- Symposia

- Monographs

- Handbook of standard models
  with parameters' values

- Handbook of standard methods

- Specialized journals

- Constitution for a professional
  association

- Annual colloquia

STAGE B:

EDUCATION, TRAINING   ————————→   STAGE C:

PRACTICAL APPLICATION

- Anthologies and reviews

- Textbooks

- Methods manuals

- Sample distribution maps

- Sample time series data

- Prototype data banks

- Case histories

- Regulatory protocols
  and standards

- Practical manuals

- Data banks

- Atlases

Source: Regier, 1975.

Research Board of Canada, the Transactions of the American Fisheries Society, as well as in the publications series of various states, national and international agencies. Similarly annual or more frequent meetings occur in the fora of various scientific associations and management-oriented agencies.

On the whole, the population dynamics paradigm as related to fisheries has developed without encountering strong organized resistance or intense antagonism. It is now a powerful approach where it is well matched to practical issues, but there do appear to be important kinds of fisheries problems for which it is not as effective as some alternative approaches (see Chapters 2, 7 and 8). That this might be the case seems to be an unsettling idea to some of the experts of population dynamics. Perhaps such discomfort may be explained by the realization that workers fully mobilized into solving crises don't wish to be distracted by suggestions that better approaches might be available five or ten years hence.

3.5 The Long-term Prospects

Presumably the population dynamics paradigm will continue to develop through Stage D and complete a strong self-organizing feed-back loop into Stages A through C. If this second major phase becomes too complex--as was threatened in some of computer-dominated IBP projects--then population dynamics may be spun out of the main stream of fisheries science and become an academic curiosity. Emphasis on exquisite detail has been proposed as a symptom of obsolescence.

Alternatively the population dynamics approach may remain unchanged in essence but be recycled as a brand new model. Thus powerful mathematical and computer techniques may get us there faster in greater comfort and style. This seems to be the current trend.

It is possible that another quite different management approach--that fish flesh be largely macerated and reconstituted into processed form--will simply displace population dynamics. That approach may be seen as advantageous to one or more major fishing nations or to large international fish markets. In this case a species-by-species accounting may become superfluous (see Regier and McCracken 1975 for further comments).

A fourth possibility is that population dynamics may be assimilated as a subcomponent into a more comprehensive model and related methods. Or the models may become generalized, with ecologically similar taxa lumped together as a single ecotaxon. The latter could be readily approached using a Schaefer model as a first step (see ACMRR 1975).

Equally other events may intervene. About the only safe prediction is that fish population dynamics as a field of study will change rapidly.

4

ecology's many variations on one theme

4.1   Ecology's Theme

The ecologist has a characteristic view-point. He (or she) looks at the total pattern of life in some defined habitat, in the belief that it constitutes one system. When he sees populations, of one species or another, growing or dwindling, oscillating or remaining strangely constant, he assumes that the regularities which make the pattern recognizable are due to the mutual influence which each population exercises, directly or indirectly, on all the others and all of them on their common physical environment. This net of relations is what he needs to understand, and the only assumption he can safely make is that it is a net--no mere tangle of causal chains but a field in which multiple influences are constantly at work. This is the interest and this the assumption which are slowly seeping through into the consciousness of Western man . . . (Vickers 1973).

That quote is an up-dated and modernized version of a theme sounded by Elton (1927) who was much influenced by Charles Darwin. Vickers' words, *belief, pattern, mutual, net,* and *field* bring to mind the sketch of the *mutual causal paradigm* (Maruyama 1975, see Chapter 2

above). This may be compared with a "systems analytic view" sketched next.

The new scientist . . . concentrates on structure on all levels of magnitude and complexity, and fits detail into its general framework. He discerns relationships and situations, not atomistic facts and events. By this method he can understand a lot more about a great many things than the rigorous specialist although his under-standing is somewhat more general and approximate. Yet some knowledge of connected complexity is preferable even to a more detailed knowledge of atomized simplicity, if (we perceive that) it is connected complexity with which we are surrounded in nature and of which we ourselves are a part. If this is the case, to have an adequate grasp of reality we must look at things as systems, with properties and structures of their own. Systems of various kinds can then be compared, their relationships within still larger systems defined, and a general context established. If we are to understand what we are, and what we are faced within the social and the natural world, evolving a general theory of systems is imperative. . . . There is . . a science of systems as such - the General Systems

Theory developed by von Bertalanffy and his collaborators. (Laszlo 1972).

Prior to his broader interests in General Systems Theory (GST), von Bertalanffy (1962) proposed an "organismic" approach in biology. Returning to Maruyama (1975, see Chapter 2 above), we note that he classifies the organismic approach with the *one-way causal paradigm.* Perhaps it is von Bertalanffy's perception of strong controlling forces acting from higher levels to lower levels in an organized hierarchy that Maruyama has used as an important criterion of classification. But in some ways GST seems intermediate between Maruyama's extremes.

The issues here addressed quite abstractly using concepts by von Bertalanffy, Laszlo, Vickers and Maruyama have seldom been addressed by ecologists either in the abstract or the concrete. Nevertheless one can detect a low level but general discomfort within ecology that may relate to a mismatch of concepts and methods. Ecologists' presuppositions appear to border on the holistic in the sense of the mutual causal paradigm, their methods are predominantly reductionist, i.e. bordering on the one-way causal paradigm. The conceptual role of the structured hierarchy, that plays such a prominent role in GST and systems analytic methods, may be to serve as a pragmatic compromise or bridge between an internal mismatch of concepts and methods.

Whether or not the above speculation has merit, the science of ecology is not well defined. No Newton has dictated a sufficient set of broad terms of reference within which proper ecological science should be done.

Popper (1972, p. 267 and 270) has interesting points:

Darwin's discovery of the theory of natural selection has often been compared to Newton's discovery of the theory of gravitation. This is a mistake, Newton formulated a set of universal laws intended to describe the interaction, and consequent behaviour, of the physical universe. Darwin's theory of evolution proposed no such universal laws. There are no Darwinian laws of evolution.

The theory of natural selection is a historical one: it constructs a situation and shows that, given that situation, these things whose existence we wish to explain are indeed likely to happen.

To put it more precisely, Darwin's theory is a generalized historical explanation. This means that the situation is supposed to be typical rather than unique. Thus it may be possible to construct at times a simplified model of the situation.

Various systems of human physical and mental health care have been developed to a level of broad practical utility in the absence of unique, basic quantitative relationships. The paradigm may be characterized as focussing on stimulus-response interactions within homeostatic systems. A similar approach has long been applied ecologically in traditional folk cultures around the world. It is easy to scoff at various claims and practices of different kinds of health care professionals and environmental and resource managers, particularly where various practitioners addressing the same issue give quite different advice. Yet no more modern scientific approach has displaced them.

If indeed the academic ecology tradition in western science is schizophrenic with respect to concepts and methods, so be it. I leave it here as a query for possible future consideration. That theme--of schizophrenia--is not of primary interest. Ecology as currently evolving is unlikely to fall apart due to possible internal contradictions--if these do exist, some way will be found of resolving or rationalizing the logical difficulties.

Here the analysis of the theme of ecology will start with an hierarchic model: organism, population, and higher levels. Parenthetically, the "higher levels" include the "taxocene" connoting an interacting set of species populations usually related quite clearly phyletically, e.g. all members of the order Pisces, or more narrowly all members of the family Cichlidae (see Section 3.3.5), or still more narrowly the *Haplochromis* species swarm in an African lake or the *Coregonus* swarm of Lake Michigan in the 19th century.

Though I seek to address a context at the conceptual level of renewable resource science and the natural environment, the more detailed discussion here relates to aquatic resources and eventually fish. I believe that comparable breakdowns could be developed using other areas of environmental resources, including the major terrestrial types.

With very notable exceptions, "fish biologists" have tended to specialize at one particular level of organization--most at the organism, some at the population, and fewer at the community level. Within a particular level one can arrange different workers along a continuum from the purely abiotic to the biotic. Some deal with the physics, hydrology, or chemistry of the habitat and treat fish at whatever level, mainly as responding mechanisms. Others focus on the fish and treat the environment, sometimes including other aquatic biota, as a passive substrate to be organized and structured by the fish. More workers fall somewhere between these extremes. Sometimes the choice is made rationally, often it appears to be influenced in large measure by the biases within disciplines to which the worker happened to have been subjected as a student.

Many fish ecologists have specialized further to a particular taxon or to a small set of closely related taxa, within one habitat or a small set of physically similar habitats. Personal sentiment or aesthetic attraction have sometimes determined the subject for specialization. Overall, this has led to a somewhat haphazard selection of taxa and waters which have happened to have received special attention; a result is that data on various taxa are patchy and difficult to analyze for generality (Loftus and Regier 1972).

4.2   Ecology at the Organism Level

The highest level of organization addressed explicitly by morphologists, developmentalists (embryologists), physiologists and ethologists is the whole organism. In North America and Western Europe, fish ecology has developed close links only with physiology. In Eastern Europe a set of generalizations have emerged on morphological and embryological ecology from the works of De Bear, Shmalhausen, Vasnecow and Kryzhanowsky.

From progress to date it appears that a well-trained fish biologist should be able to deduce a wide variety of information, on a taxon's ecological role and its preferred habitat, from an examination of gross internal and external morphology. This source of information is generally ignored and fish ecologists undertake costly sampling programs to learn with painful slowness what should be obvious if the myriad limited inferences on organism form and function had been generalized effectively.

The finer details of external morphology, and particularly of coloration, have particular relevance in ethology. Spawning phenomena have interested ethologists and it is perhaps with respect to spawning that one might first expect effective collaboration between ethologists, ecologists and perhaps physiologists. Advances have been registered from an ecological viewpoint, particularly by such Swedish workers as Fabricius, Lindroth, and Gustafson.

In physiological ecology the work in "physical factors" is well known. Fry's (1947) formulation apparently acted as a more recent stimulus in a tradition that goes back at least to Liebig's "law of the minimum" (Fry 1971). The magnitude of separate and joint effects of temperature, oxygen and carbon dioxide concentrations, and other factors on fish activity and well-being have been measured and modelled for a number of freshwater species. Some generalizations over different taxa have been proposed (Zahn 1962; Brett 1971). Fry's contributions to all this have recently been reviewed (Lawrie and Kerr 1976).

Inferences and estimates from laboratory studies on physical factors have been checked with field observations (Ferguson 1958) and are now being field tested with encouraging results. This approach is being applied to the practical problem of predicting effects of "thermal pollution" on fish distribution. Also fish cultural establishments are planned to optimize physical environments for the fish.

The potential usefulness of inferences on "physical factors" for predicting a species' seasonal station and movements in its habitat has hardly been developed at all. Such physical variables as temperature, oxygen concentrations, and light intensity gradients are more easily measured and monitored than is fish distribution itself. It follows that such inexpensive data, together with tested general inferences on physical factor preferences, could and should be used to predict where in the habitat, and when in time, sampling for a particular taxon would be most efficient. This is now done informally; it should be approached scientifically and explicitly. Thus inferences at the organism level would provide an improved basis for efficient sampling at the population level.

A second large group of physiological ecologists are measuring the toxicity of a wide variety of cultural wastes (pollutants) and biocides, particularly in the U.S.A. and U.K. A number of taxa have in effect been selected for bioassaying mixed effluents in pollution control protocols - their physiological responses to various toxic materials singly or in combination having been measured. The experimental and statistical methods, and the mathematical formulations of models on pollution physiology closely resemble those used in physical factors work. Inferences from the two groups are readily interrelated (Alderdice 1971). Alabaster et al. (1972) have empirically measured the relationship between estimates of bioassay parameters obtained in the laboratory and the actual longterm responses of the relevant species in polluted waters.

Thirdly, much fish disease and parasite work has been essentially at the organism level. It too has been applied practically, e.g. by Schäperclaus in Germany, and by

Lucky and Dyk in Czechoslovakia with respect to pond fish production.

It appears then that work on the physiological ecology of fish exhibits a healthy balance between laboratory and field work, and is maturing to the point where the science is becoming very useful on practical questions (Regier 1974).

4.3   Ecology at the Population Level

The study of the evolution and taxonomy of fish has been far from exhausted. The Linnean paradigm is under attack as being too constraining at the "species" level, differences between Darwin and Wallace are unresolved, etc. Despite creative work by Svardson and others, the salmonines and coregonines continue to baffle us. Hundreds of "species" in African cichlid swarms remain to be identified and their evolution rationalized. Sympatry and allopatry defined rigidly are seen as unrealistic extremes of a probabilistic continuum, but much on-going taxonomic work with fish still implicitly treats that dichotomy as operationally valid. Ecologists reared in the dogma that Linnean species are basic to all biology risk schizophrenia when they encounter taxonomic swarms. More functional approaches, from an ecological viewpoint, are being considered or are simply being implemented with hardly an apology to the classical taxonomists. Elton, Lindeman and many others have done so, particularly when addressing events at the community or ecosystem level. Statistical and modelling techniques are being developed further by which a worker can derive a functional ecological taxonomy more or less objectively.

The tendency in classical taxonomy is to emphasize differences and distinctives.

This is a useful approach in the study of population genetics, but an overemphasis on differences, or a mindset that cannot overlook such differences in order to achieve another purpose, may hinder the search for ecological generalizations.

Work in population genetics with fish is closely related to ecology. It is applied to pond fish culture, particularly with the carps. As demands for protein intensify, such work will rapidly expand to many species worldwide.

Hybrids have long attracted attention. For the Laurentian Great Lakes a hybrid char, the splake, was developed through extensive selective breeding to accept an ecological role in which it would be affected only minimally by the parasitic sea lamprey--a recent invader. For the recreational fishery in the United States, various small centrarchids have recently been hydridized to provide the taxa with "hybrid vigour" in growth and survival. Cichlids are being hybridized in the hope of producing unisex offspring for planting ponds; otherwise reproduction tends to be too successful with consequent stunting, as with small centrarchids.

Stimulated by cereal agriculturists, some recent international interest in conserving "gene pools" has been welcomed by some fisheries workers. The extinction of a number of taxa, particularly among the salmonids, is regretted. Programs are planned to prevent direct extinction, perhaps by creating reserves, and also to prevent swamping through hybridization of remnants of some taxa through the introduction of closely related forms. Earlier compendia by Berg, Banarescu, and others need to be updated, refined, and extended to other areas.

At the population level, disease and parasite work has involved life history aspects of the pathogen usually investigated by specialists most interested in taxonomy or in control of the organism. By comparison, the *ecological* effects on the host population have seldom been the focus of a major study, except perhaps in Poland by Klekowski and others, and in Canada in the work of Miller and others on *Triaenophorus* infections. Also in Canada, Dechtiar has for over a decade regularly monitored the relative infestation of a number of fish taxa by quite a number of pathogens in the Laurentian Great Lakes, but those data remain to be analyzed from a broad ecological viewpoint. Evidence suggests that certain pathogenic organisms may on balance benefit from organic pollution, and the interest in this subject is likely to grow.

A field ecologist intending to study fish is faced at the outset with the task of locating these creatures in an alien and largely opaque environment. Many works continue to be published on the distribution and movement of a particular taxon's cohorts from birth to death. Such space-time trajectory, aside from physical factor constraints, generally has its annual and daily rhythms, with other major components related to changing characteristics of the cohort with age. The trajectories of only few taxa have been closely described and then generally only with respect to particular waters. If, as has usually happened, these particular waters have been subjected to increasing levels of cultural stress, then a space-time trajectory inferred at one stress level may have little predictive value at a higher or a lower stress level, even within the same water. Obviously some general inferences would be very helpful.

Much of the recent practical fisheries management of wild or common-property resources has been based on study of population dynamics (see Chapter 3). With freshwater stocks the theoretical and practical development has been due originally to Baranov, Ricker, Fry, and others and to statisticians such as Chapman, Robson, and Paloheimo. Large benthic stocks and also pelagic species of the seas have been studied using a methodology broadly similar to that used with freshwater population dynamics, though the three differ from each other in interesting ways that need not concern us here (see Gulland 1972).

With measures of growth rate, "natural" (non-fishing) mortality rates and the appropriate mathematical model, one can predict the maximum yield to be derived from a cohort, assuming in the simplest case that recruitment of subsequent cohorts is very largely independent of fishery practices. If heavy fishing is allowed too early in the life-span of a cohort, such overfishing will preclude a maximum harvest. This type of overfishing is sometimes referred to as "economic or growth overfishing" in that recruitment is not threatened and there is rapid recovery of fish and fishery after a reduction in fishing intensity. The main concern is with low yields, hence is "economic", rather than with the biological effects on the stocks. If the nature of the fishery process and the ecology of the fish resource is such that recruitment is severely impaired, then "recruitment or biological overfishing" is said to occur (Ricker 1961).

Population dynamics theory as presently elaborated is practically useful with respect to growth overfishing assuming that the common property--open access-- willing consent impasse is not operative (see Chapter 9). As a conventional fishery develops, from a beginning of no fishing at all, growth overfishing is likely to occur at relatively moderate fishing stress. With increasing stress, the initial presuppositions on the nature of the fish population responses become less realistic. Extensive attempts to incorporate explicit stock-recruitment components into the models of population dynamics have only been successful with some anadromous salmonines. Even in the latter case it is not clear how such information is being incorporated into the decisions on the level at which the harvest is to be controlled.

Eipper and associates (1974) have used the methods of population dynamics to measure average production (in kg/ha/yr) of harvestable fish of salmonines and centrarchids in U.S. recreational fishing ponds of less than a hectare in area. These studies are of methodological interest in that numerous replicates, over ponds and years, were analyzed. Variances of estimates of average production could thus be estimated nonparametrically. In population studies of a particular taxon in a specified water, variances generally can only be estimated parametrically, i.e. following acceptance of a series of assumptions on the probabilistic behavior of the organisms and the sampling devices. The effect due to the special characteristics of the body of water generally cannot be measured directly. For purposes of generalizing over different populations of a taxon, most population dynamics studies are in effect restricted to samples of size one.

From a consideration of the full range of population and community responses, to exploitation and other stresses, population dynamics methods are likely to be efficient over a relatively limited range of low to moderate stress levels. In such a context populations dynamics may seem to be a misnomer, and "population statics" might be used in its place (see Chapter 6).

With respect to details of the models, a standard assumption is that $dN/dt = -(F + M)N$, where N is number, t is time, F is the fishing mortality coefficient and M is the "natural" or other-than-fishing mortality coefficient. Simply the assumption that F and M are fully independent would lead one to suspect that the model cannot extend over a wide range of F or M or both. Yet the degree of unreality with respect to this assumption has seldom attracted interest; generally it appears to be assumed that problems associated with recruitment are more critical whether or not an explicit stock-recruitment component is incorporated.

To the extent that fisheries have actually been developed and managed using these concepts, it has generally been with greatly simplified versions of the models. Or the variables such as growth, mortality and recruitment may not even have been assembled explicitly into a single comprehensive model. An empirical method for managing fish described by Abrosov (1969) of the U.S.S.R. may well be more robust and practically useful than a sophisticated analysis of population dynamics, at least on the small to medium-sized fisheries. Similarly in marine work, some of the more successful management attempts were implemented following analyses using very simple, "stripped-down" models, several of which are due to Gulland.

Yet one can hope that man's demands on fishery resources will eventually be controlled so as to prevent overfishing to such low levels that there is no benefit to human users. An eventual quasi-steady state will again provide scope for the fine-tuning possibilities within present population dynamics theory.

Several attempts at modelling the interactions of fish species have been successful, from a theoretical viewpoint. Larkin's (1971) work with Pacific Coast salmonines has used classical population dynamics models with a stock-recruitment component. Forney (1971, 1976), has related the dynamics of two percids to each other. Nilsson (1967) has treated dynamics rather more implicitly in his work on the interactive segregation between two or three salmonid taxa in various Scandinavian lakes. Work such as this is a step in the direction of community dynamics from "simple" population dynamics.

Turning to sampling methodology, as already hinted in the discussions on morphology, physiology, and space-time trajectories of cohorts, the theory and practice of sampling fish populations especially in large waters has never been treated comprehensively. When we note that the other half of any sampling problem--the operational procedures and "calibration" of the sampling gears--is in similar disarray, we might justifiably term the situation scandalous. To measure the sampling characteristics of fish gear has generally been seen by the biologists as an unpleasant and unsatisfying experience. In much of science these duties are absorbed by the industrial manufacturer of sampling and testing equipment. The manufacturers of fishing gear also know certain of the sampling characteristics of their products, but usually not in sufficient detail for purposes of ecological study. Thus they may know quite well what minimum size will be captured by gear of a particular mesh, but not know the relative probability of capture as a function of size above that minimum size. The problems are not simple as recent studies with gillnets clearly show (Hamley 1975).

In smaller waters the use of electric shocking gear, fish toxicants, seines and traps are quite effective. But again the gear is seldom "calibrated" objectively, nor a standard methodology specified in sufficient detail to ensure that sampling is reproducible to a satisfactory degree (Regier 1975).

4.4 Ecology at Taxocene, Community and Ecosystem Levels

The state of the science of aquatic ecology with respect to higher levels of organization and its practical applicability to a range of issues have been documented extensively by a recent international working party (ACMRR 1976). Table 4.1 is taken from that report and illustrates a rich variety of approaches now underway.

Of the ten pairs of traditions listed, the first three pairs have been developed to a useful state and are incorporated widely in practice by government agencies. With respect to the final six pairs, practical application has occurred in some parts of the world--sometimes only on a trial basis; some will doubtless be generalized and rendered operational during the next decade.

4.5 Many Variations

The evidence shows that aquatic ecologists are pursuing a wealth of ideas and concepts many of which have been applied here and there to renewable resource problems. Yet, relatively few practicing ecologists are conversant with more than a few of them. Most ecologists are quite parochial in their roles as scientists, as are most fisheries scientists.

From the viewpoint of interdisciplinary collaboration it is encouraging to note

Table 4.1 ECOLOGICAL TRADITIONS WITHIN STUDIES OF AQUATIC COMMUNITIES AND ECOSYSTEMS AS VIEWED IN A THEORETICAL RATHER THAN A PRACTICAL ISSUE-ORIENTED VIEWPOINT.

	Structure	Function--Processes
1.	Whole system classification biotopic typology: Thienemann, Hutchinson, Müller, Moyle, T.A. Stephenson, G. Thorson	Biogeographical processes: Lindsey, MacArthur, Svärdson, Magnuson
2.	Whole system abiotic structure, e.g., "morphoedaphics": Rawson, Ryder, Jenkins, Huet, Cuinat	Whole system responses, e.g., "succession" due to sediment in filling of basin: Vollenweider, Rigler, Harvey, Sheldon Sutcliffe, Regier, Lewis
3.	Indicators--species and subsystems: Saprobensystem: sanitary engineers, Reich	Population dynamics; eutrophication, oligotrophication, and pollution studies: McElreath, Johnson, Brinkhurst, Schindler, Thomas
4.	Niche structure: Keast, MacArthur	Natural selection, e.g., interactive segregation: Nilsson, Dodson
5.	Trophic pyramids or gross networks: Ryther, Cushing, E.P. Odum	Productivity coefficients and ratios: community metabolism and respiration H.T. Odum, H. Welch
6.	Complicated food webs: Isaacs, Tyler	Trophodynamics, energy and material transfer coefficients: IBP studies, Steele, Dickie, Parsons
7.	"Balance," biomass ratios: Swingle, Anderson	Interactions between taxonomic groups: Brooks, Hrbáček, Shapiro
8.	Pattern distribution and species associations: W. Stephenson	Species interactions: Lewis, Paine, Connell, Gallucci
9.	Diversity/complexity: Pielou, Sanders, MacArthur, Cairns	Stability/resilience: Holling/Jones, Margalef
10.	Particle size profiles: Sheldon	Particle size dynamics: Kerr

Source: ACMRR, 1976.

that Rapport (see Regier and Rapport 1976) has identified quite a number of conceptual commonalities between economics and ecology, and Table 4.2 indicates some of these.

The conceptual framework for the preceding sections of this chapter is quite conventional to most fisheries workers, I expect. Other than the hierarchic feature and the initial focus on anatomy,

Table 4.2   SOME COMMONALITIES BETWEEN MICROECONOMIC AND ECOLOGICAL THEORY.

Topic	Economics	Ecology
Consumer choice . . .	Theory of consumer equilibrium . . .	Optimal foraging theory (Rapport, Schoener, Charnov, Covich, Tullock)
Production . . .	Investment theory . . .	Life history strategies (Schaffer) Parental investment (Trivers)
	Production theory . . .	Foraging and population growth (Rapport and Turner)
	Contract theory . . .	Energetics of bumblebees (Oster)
	Optimal input mix . . .	Social caste system (Wilson)
	Location theory . . .	Central place foraging (Hamilton and Watt)
Consumer-producer . . . interactions	Theory of markets . . .	Community interaction (Rapport and Turner) Predator-prey interactions (Holling)
Competition . . .	Oligopoly theory . . .	Interspecies competition (McArthur, Rapport and Turner)

Source: Rapport and Turner 1975a, b, 1976; Rapport, 1971

physiology, ecology and evolution, the framework is not well defined. A less conventional but more definitive approach has been sketched by Regier and Rapport (1977) and may be more useful than the conventional for understanding contemporary ecology. Further it appears to us that each of the five major approaches sketched below is in fact shared with other disciplines such as geography, engineering, agriculture, forestry, economics and political science.

a)   Phenomenological, spatial, temporal descriptions — Under this heading may be subsumed the nature study accounts stimulated by rather romantic perspectives on nature, accounts of travels by explorers, classificatory work by biogeographers, by taxonomists and by observers of annual cycles in a natural community, etc. "Surveys" of various kinds still form a major part of the budget of ecological information services. The data are sometimes not "worked" beyond the stage necessary to summarize them for ready reference in the form of spatial maps, temporal time series, or classificatory tables. More commonly now inferred relationships are summarized in the form of empiric mathematical relationships fitted statistically.

b)   Stimulus-response processes in homeostatic systems — This school seems to have evolved largely from the practical interests of (1): the managers of renewable resources and the natural environment; (2) scientists concerned with production of economic goods through agriculture, forestry, fisheries, wildlife, and rangelands; and (3) scientists seeking to control or mitigate pollution and other deleterious stresses or abuses of the natural environment. The models frequently relate the response of some variable (of interest or value to the user) to an increase or decrease of some other variable (over which the user can exercise control or with respect to which he can take appropriate accommodative action). The magnitude, direction and time characteristics of the response are usually of primary concern. Perhaps because much of the science of the stimulus-response school was developed for practical purposes in close association with practical people relatively little of it is included in academic textbooks of ecology.

c) Natural selection and resource allocation — The time parameters of evolutionary processes are so slow as to make it worthwhile on occasion to treat ecological processes, which are often much faster, as being in a different class. But what an organism can do in its habitat is strongly constrained in the short run by its hereditary script. An understanding of what those limits are and how quickly they can change - if at all - yields information, for example, on the species composition to be expected in an ecosystem following a specified impact on it. A study of resource allocation, niche overlap and other species interactions has its counterpart studies in the field of economics which is not surprising since the study of social economic systems and that of evolutionary ecological systems proceeded through mutal exchange of models especially in Britain during the past two centuries.

d) Compartment-flow analysis — During the past two decades much effort has been directed toward an analytical description of ecological systems. Quasi-discrete compartments or components are identified and characterized, the flows of energy and/or materials in various forms are then measured. Various postulated controlling mechanisms are modelled, the whole process may be computerized. The model may subsequently be exercised to discover if the output is reasonable and, if so, the model may be extended incrementally with respect to a number of variables to provide forecasts of what to expect from proposed impacts of specified types. The second generation of such attempts have now been completed and are not yet of much direct use to practical people but the third generation attempts now beginning may be quite useful.

e) Hierarchic general systems approaches — Some rather philosophically-inclined theorists (to be distinguished from the "applied systems analysts") have sought to identify the properties of living systems that control the degree of homeostasis that such systems exhibit--to the observer predisposed to perceive homeostasis. These attempts usually begin with a presupposition that reality is organized hierachically. Hierarchy and homeostasis permeate the thought of most ecologists to the extent that few question the reasonableness of these assumptions or whether some alternative assumptions might not also lead to a satisfying understanding of ecological systems. But the general systems theorists do address such issues directly and seek to learn the consequences of these assumptions and other features of broad ecological models.

5
an emerging methodology for major issues

5.1 Big Changes Underway in Science

Is it the fate of fisheries scientists simply to be swept along or are they exerting some influence on the future of science. Consider 1985:

a) Will research grants to University fishery scientists be made available under much the same rules as today? Fifteen years ago, did U.S. scientists foresee the Sea Grant Program? Will some innovation of equal or greater importance come along before 1985? If so, what will be its goals, terms of reference, and operating guidelines?

b) What will be the emphases in research of government agencies with responsibilities for renewable resources and the natural environment? Several years ago The Institute of Ecology was invited to review EPA's NERC (Corvallis) network. The review team found that relatively little attention was being directed to events at ecosystem or higher levels of organization. Will a marked change occur along these lines?

In Chapter 1 some findings of the 1975 Advisory Committee on Marine Resources Research (ACMRR 1975) were quoted. They might be paraphrased thus: for the immediate future we will need to work faster with methods that we now understand, but we should look for alternatives that may be more appropriate to the changing times. What alternative approaches are going to win in the inevitable competition?

c) What kind of research contracts will become available in the private sector or to the privately incorporated researcher? Presumably the environmental impact assessment procedures will evolve rapidly--mostly they are now at a quite primitive stage of development. What will they be like at maturity? A number of symposia have recently addressed the question without resolving it.

What will be different about 1985? Many observers of research trends judge that significantly more of the overall scientific and research budget will be allotted to interdisciplinary work.

In October 1975, the University of Tennessee, with NSF funding, sponsored a conference on "the management of large-scale interdisciplinary research in universities". There it was argued that "if universities ignore or resist interdisciplinary research, private industry will dominate the field" (T. Owen 1975, personal communication).

Yet there are great *academic scientific* challenges implicit in the trend to interdisciplinary research if by this term is meant "joint research conducted by people from different research areas where the final product tends to obscure the individual input, thus emphasizing the importance of the synthesis of the various research areas" (T. Owen, 1975, personal communication).

At about this point in the proceedings, a red herring is usually whisked across the path along which the discussion is going. "Pure research will be starved for funds, will wither away, and with it will go the creative source of future scientific progress". This is a red herring because it quickly exposes sensibilities and raises emotions, and because the issue of pure versus applied is not necessarily closely related to the growing interest in interdisciplinary science and technology.

5.2 Disciplinarities

First consider a possible set of meanings for the various prefixes that are coming to be attached to the term "disciplinarity". Figure 5.1 is taken from Jantsch (1972, p. 223). The first order approximations to the definitions stand out rather clearly in this graphic

presentation. Note in particular the conceptual, academic and scientific challenges implicit at four different levels:

Disciplinarity--to specialize in a well-organized field; at worst to learn more and more about less and less.

Crossdisciplinarity--to expand an approach to exercise broader conceptual leadership; at worst to create a cross-disciplinary hegemony by coercive means.

Interdisciplinarity--to clarify higher-level "syntheses" or transcendent concepts; at worst to be satisfied with terms that have no realities corresponding to them (cf. nominalism).

Transdisciplinarity--to participate in "rational creative action"; at worst to become like a billard ball with only one fate in view, i.e. to drop out of sight.

Parenthetically the negative echo attached to each of the four terms is consistent with Garrett Hardin's espousal of "pejorism" (Hardin 1976). Pejorism presupposes that if something can go wrong then it probably will, as contrasted to "meliorism" which begins with rosy presuppositions. Of course, the pejorist seeks precautionary measures to forestall utter failure.

Hence, knowing the likely high-entropy end points of these various initiatives, if allowed to proceed conventionally, we should be able to develop structures and processes to reverse the processes of entropy growth, not only momentarily but for extended periods.

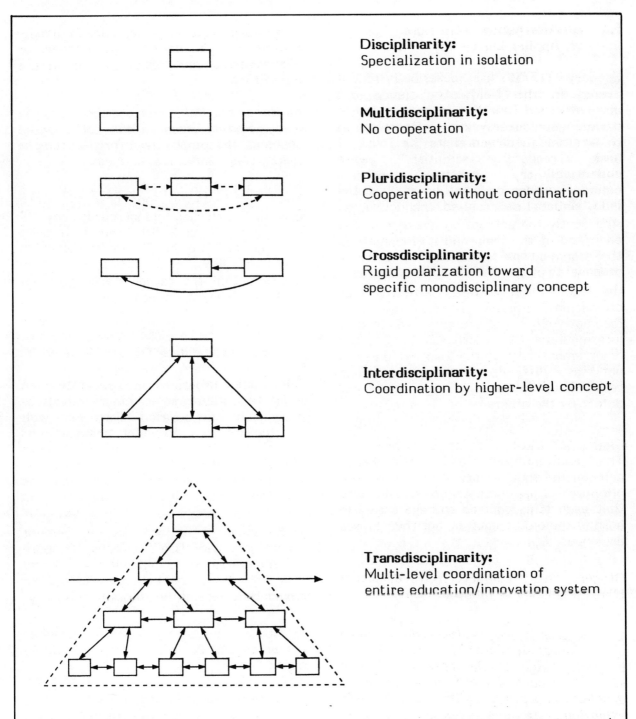

Disciplinarity:
Specialization in isolation

Multidisciplinarity:
No cooperation

Pluridisciplinarity:
Cooperation without coordination

Crossdisciplinarity:
Rigid polarization toward
specific monodisciplinary concept

Interdisciplinarity:
Coordination by higher-level concept

Transdisciplinarity:
Multi-level coordination of
entire education/innovation system

Figure 5.1   Steps toward increasing cooperation and coordination in the Science/
Education/Innovation System (Jantsch, 1972).

5.3 An Alternative to the Pure vs. Applied Dichotomy

Gregory (1975) has described recent events in the field of "science and discovery in France". In the preceding decade a controversy had run its course as to whether fundamentalists (i.e. pure or basic research scientists) were collectively the locomotive of the train called Progress or whether it was the industrialists (i.e. applied technologists). Apparently the perception has grown that each end of the train had its locomotive; the argument as formulated by extreme proponents of the two viewpoints obscured the reality that a fairly effective co-ordination already linked the two locomotives. Focussing on that intermediary process, the complementarity of the two extremes is not as difficult to perceive and rationalize, thus neither side need fear defeat by the other.

Figure 5.2 illustrates Gregory's proposal. The analysis-based and action-based components approximate in concept to the old pure vs. applied categories. But note that each is recognized to have a role in both of the old categories, but their biases are opposed.

Gregory has characterized effective transfer science as follows:

a) a transfer science is always downstream of an important sector of activity. It is, in fact, owing to the pressure of this sector that each of these sciences is born, develops, and continues to exist. If their evolution is examined, it is seen that their importance and vitality are closely linked to the prosperity of the upstream sector from which they emanate and the rapidity of its evolution;

b) transfer science must remain deeply rooted in high quality basic science, or else sclerosis will rapidly set in in the laboratory;

c) no direct linking is possible between analysis-based sciences and action-based sciences, (hence the need for the transfer link).

These concepts are not new and were current in fisheries science circles several years ago (Regier 1974). They have been incorporated into recent research planning (see Chapter 9). If accepted, they would get us out of the rather sterile pure vs. applied setting of the past.

5.4 Some Candidate Transfer Science

Some major programs have recently been instituted and/or some older initiatives have geen reinvigorated. Each may well develop its own "transfer science" and many of them will intersect with "fisheries science" as it may be conventionally perceived. Consider the following possibilities:

a) Establish national management sovereignty over fish resources off shore 200 miles with some collaborative control over migratory anadromous and marine stocks that exceed the 200-mile boundary.

b) Restore and/or rehabilitate renewable resources and environmental quality, say of the Great Lakes.

c) In large construction projects or new social programs, seek to internalize the costs, that were once externalized routinely, through impact assessment procedures and related controls and regulations.

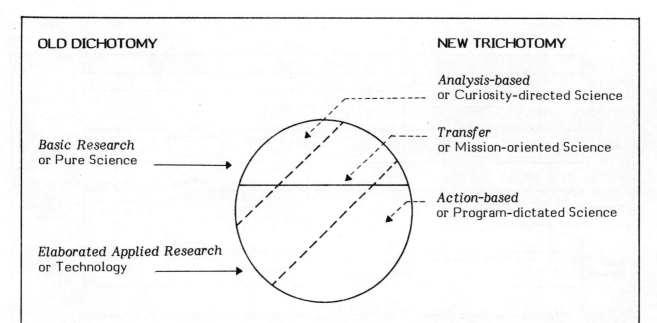

Figure 5.2   A conceptualization of science based on French experience that transcends the basic vs. applied dichotomy. Italicized terms are those used by Gregory (1975); other terms were used by Regier (1974) who made a similar proposal for fisheries work. Both are broadly congruent with some proposals by OECD (Regier and McCracken, 1975).

d) Recover resource rent from common property resources, necessitating solution of the "common property--open-access-- willing consent" puzzle.

e) Develop a system of national environmental accounts which will involve difficult choices of what should be monitored. (Why not certain fish stocks and aspects of their ecosystems?)

f) Through regional land use zoning and related planning seek to accommodate conflicting interests each of which is deemed important to society's welfare.

g) In order to move toward a "conserver society", develop a balanced process of technology assessment and programs to expedite control of waste, material re- cycling, design of more efficient structures, etc.

h) Accommodate more of the demands of native peoples and some other apparently discordant themes such as environmental activism.

For each of these examples, I expect further mobilization to occur leading to clearer definition and perhaps a "transfer science". Fisheries scientists will be mobilized into these activities, which may well be organized in an interdisciplinary and eventually a transdisciplinary mode. What will be the feedback to fisheries science? Is there anything we can do now to ensure that fisheries science will receive due attention in the collaborative interdisciplinary process? The last five chapters of this book try to deal with those kinds of questions. But first a brief glance at a couple of proposals on the methods for achieving such transfer science.

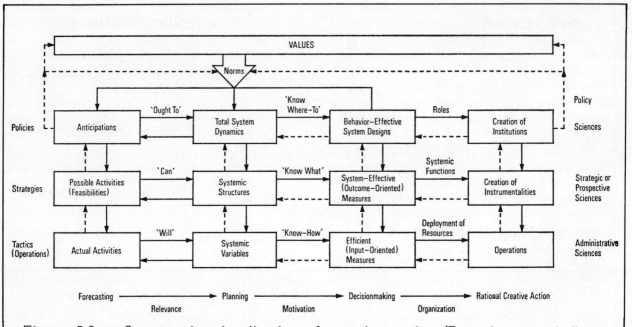

Figure 5.3 Structured rationalization of creative action (Dotted arrows indicate feedback) (Jantsch, 1972).

5.5 Structured Rationalization of Creative Action

Figure 5.3 (Jantsch 1972, Fig. 2.3) emphasizes structure, or so it would seem at first glance. But in his text, Jantsch focusses on the arrows, particularly the role of feedback. And in specifying "norms" as the transformation between widely held values and directed policies, he disassociates himself from any emphasis on an overall process that would specify explicit goals and then mobilize the political process to achieve them. Hence his approach presupposes evolutionary changes acting back on the norms and also the values. Nevertheless, Jantsch sees the predominant flow of authority and action stimulus from the top left to the bottom right, with the process gradually coalescing toward the bottom right.

Jantsch has used the same format to sketch the characteristics of certain primitive or degenerate types of social decision-making, e.g. bureaucratic, technocratic, and futurocratic. If I understand correctly, the major Canadian fisheries agencies have recently graduated from the bureaucratic to the technocratic type and have some distance to go to meet Jantsch's ideal of creative action. Perhaps they will never get there.

Whatever the future, most explicit attention is now directed to the *four* boxes on the bottom right. Here, at the national level, *instrumentalities* are agencies like the Fisheries and Marine Service and the Environmental Protection Service in Canada and NMFS and EPA in the United States, while *operations* involve their major programs. In Canada relatively little public or governmental attention is focussed on the four boxes to the upper left. There is quite a lot of general discussion about values but not about norms.

Pritchard (1976) has used this model by Jantsch as a convenient framework by which to analyze the current state of understanding of the "aquaculture option" in Canada. In a somewhat similar way it could be used with each one of the eight issues of the preceding section (5.4 above). Whether it is worthwhile to take such a comprehensive approach to these kinds of issues is open to question. Some persons seem to feel that it isn't, and propose instead a less structured "field" or "creative learning" approach (Chevalier and Burns 1975; Fox 1975).

5.6 Field Approaches and Interdisciplinarity

Raiffa's (1973) proposed methodology for the International Institute of Applied Systems Analysis (IIASA) may be viewed as a type of "field approach":

ASA .. involves the use of techniques, concepts and a scientific, systematic approach to the solution of complex problems. It is a framework of thought designed to help decision makers choose a desirable (or in some cases a "best") course of action. The approach may entail such steps as

a) recognizing the existence of a problem or a constellation of interconnected problems worthy of and amenable to analysis;

b) defining and bounding the extent of the problem area. It is necessary on the one hand to simplify the problems to the point of analytic tractability and on the other to preserve all vital

aspects affected by various possible solutions. The difficult judgement upon the inclusion or exclusion of problem elements--balancing their relevance to the analytical grasp of the situation against their contributions to manageable complication--often determines the success of systems research;

c) identifying a hierarchy of goals and objectives and examining value tradeoffs;

d) creatively generating appropriate alternatives for examination;

e) modelling the . . . interrelationships among various facets of the problem;

f) evaluating the potential courses of action and investigating the sensitivity of the results to the assumptions made and to facets of the problem excluded from the formal analysis;

g) implementing the results of the analysis.

Precisely because ASA is a rational approach rather than a technique, the list of steps above should be understood in a qualified sense. Not all the steps need be included in every instance of responsible system analysis. Some steps may be handled in a more formal manner than others . . . (Raiffa 1973)

The question of what is an efficient and workable interdisciplinary methodology is currently under study by numerous groups. Though efforts to achieve inter-disciplinarity were made with UNESCO's Man and the Biosphere (MAB) program, some ecologists preferred to play it safe in a crossdisciplinary approach with ecology as the lead discipline. That is the way MAB may end up. The International

Institute of Applied Systems Analysis (IIASA) tends to be oriented toward computer methodology and may be equally content with crossdisciplinary or interdisciplinary modes. Perhaps the Research Management Improvement Program of the U.S. National Science Foundation will help to bring some broad order to the discussion with its final reports of 1976-7.

The Organization for Economic Cooperation and Development (OECD) through its Paris headquarters has perhaps stimulated more progress along these lines than any other oranization. During the past decade various projects of the United Nations Development Program (UNDP) have sought to work within an interdisciplinary mode; however the actual approaches in projects have seldom if ever surpassed the multidisciplinary or crossdisciplinary modes (see Figure 5.1).

The trend toward explicit interdisciplinarity seems to be one of the major current trends. The focus has often been more on methodology than on conceptual commonality or transcendence. Methods are certainly not neutral toward concepts. Much of the work along these lines appears to be consistent with the one-way causal paradigm of Maruyama (see Chapter 2). If ecological concepts are not adequately served by such a methodology, then ecologists should concern themselves to learn in which direction the current trend toward interdisciplinarity is leaning.

6

selecting an interdisciplinary approach

6.1 Perceptions

The origins and characteristics of an academic species are no more accidental than the stripes of a zebra or the neck of a giraffe. While anatomically bizarre, they (may) have high functional value in a given environment . . . it is through study of man's cognition of the man/environment relationship that further theoretical advance is to be expected. (Burton, Kates and Kirkby 1974).

For purposes of the present chapter I will assume that the common property--open access--willing consent impasse (Ostrom and Ostrom 1971) in effective management and consequently in useful research will be circumvented by some appropriate high-level political action.

The meaning of the word "interdisciplinary" as used here is consistent with that implied by Jantsch (1972) as shown in Figure 5.1. In particular I focus here on the "higher-level concept" that transcends the more conventional disciplines, as shown in Jantsch's definition of his interdisciplinarity.

6.2 A Mean-Variance Context Generalized for Fisheries Management

Both the mean and variance of a variable--say of a yield output of a natural resource system--have long been of interest to decision-makers. Extrapolated or projected into the future they measure the expectation, and the uncertainty to be associated with the expectation.

Quite a number of somewhat different scientific schools or traditions have developed with respect to biological aspects of management of fisheries in the wild. An attempt to classify these within a large-scale mean-variance context led to a two-by-two table in which both mean and variance were divided into small and large components. An alternative though roughly analogous approach was attempted following a suggestion of Beverton (1974) that continuous scales be used for the two variables (see Figure 6.1).

The horizontal scale, expectation of a unit resource, can extend from less than 10 kg/yr of bass from a small managed pond to about 10^{10} kg/yr of anchoveta in the

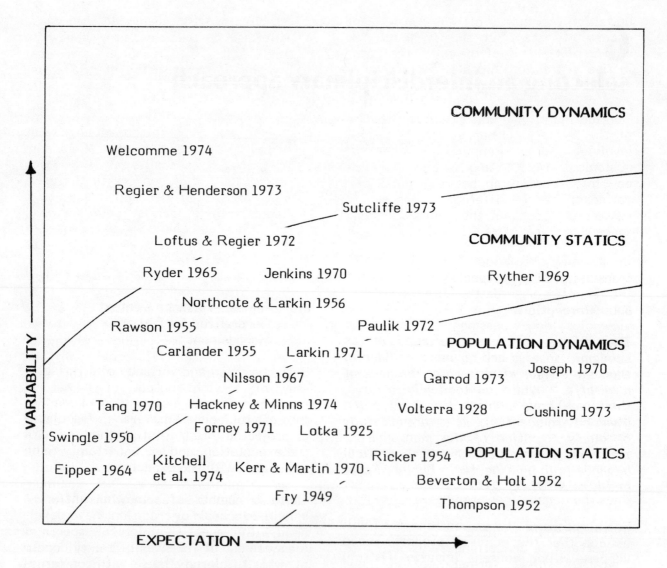

Figure 6.1   The most appropriate level of ecological organization to be addressed directly in research and management of a fishery resource depends on the average catch or expectation and the predictable uncertainty or variability of the individual unit stock or stocks comprising the resource. Both axes are logarithmic and extend over some 10 orders of magnitude, up to 10^{10} kg/yr. Each of the four alternatives, here sketched for ecological science, has roughly congruent counterparts in economics and sociology. Different kinds of decision-making rules and mechanisms have evolved for different parts within this expectation X variability domain. Unless special precautions are taken, fisheries and other stresses singly or severally tend to destabilize the resource ecologically and "drive" the resource system upward in that domain. Eventually an interdisciplinary approach, that was originally quite adequate, may be rendered obsolete through such destabilization.

upwelling system off Peru. The uncertainty, say, in the form of the standard deviation of the predicted catch in five years time has a similar range of about 10 orders of magnitude.

Considering all the unit resource fisheries of the world--whether single species or involving some lumping together of ecologically interacting species--there is no close relationship between mean and variance, hence different fisheries are scattered throughout the two-dimensional space shown in Figure 6.1.

In showing that Figure 6.1 is a useful transdisciplinary perspective I first sketch the significance of the four edges or boundaries from a decision-making viewpoint encompassing a number of disciplines. Then various schools of fisheries research are located approximately within this space and another orthogonal dimension inferred to explain the similarities and differences thus derived. Finally the perspective is shown to be equally relevant from the viewpoint of seeking to manage fisheries, to control or manage waste loading, or to regulate other uses of aquatic systems. Thus the concept has broad generality.

a) Lower side. Where year-to-year uncertainties concerning yields are low, rather simple formulations of goals and decision rules are possible. Scientific models can be made quite precise and some useful approach to optimization may be possible. Human communities relying on such "stable" resource systems are likely to be well-organized and stable--recall that common property-free access-willing consent problems have here been assumed to have been solved in order to simplify the present discussion. Close regulation of resource use, well developed social infrastructures, stable low-risk economic practices, emphasis on tradition, etc., appear to be typical of resource and user systems characteristic of the lower side of Figure 6.1.

b) Upper side. Here the converse of what was listed for the lower side appears to apply generally. The high uncertainty is of concern to the user, entrepreneur or decision maker, if it stems from a lack of understanding, since he will have no assurance of recovering any investment of time or resources directed towards harvesting such systems. Even where large fluctuations may be understood in an actuarial or causal context, one of the primary concerns of users will be to develop some harvesting strategy or institution that will provide for an averaging of the good and the lean years and still yield a profit. In extreme instances of uncertainty concern about loss of life or about bankruptcy will predominate and here a minimax decision rule may apply. Under minimax the first consideration is to minimize the risk of maximum loss rather than, say, to maximize the yield itself.

Systems of high uncertainty occur where large-scale environmental fluctuations can radically alter the nature of the resource system, where storms, floods or tidal waves can destroy harvesting facilities and claim lives, or where social institutions are inherently unstable. If a nation finds it desirable that such resource systems be harvested, it may find it necessary to establish broadly based insurance mechanisms. Thus a ravaged system may be declared a "disaster area" and relief funds may then be made freely available.

Part of the process of "development" usually involves engineering schemes to control floods and to protect harbors and coast lines from storms and tidal waves. This is obviously an attempt to reduce uncertainty by controlling nature and to the extent that it is successful the domain over which minimax procedures might apply directly is reduced.

Summarizing the contrasts between the lower and upper sides, optimization procedures have little or no relevance at the upper, but rise to achieve full relevance asymptopically to the lower side. The converse seems true for a minimax rule. To the extent that uncertainties can be modelled actuarially and risk be quantified, the role of optimization may be extended into systems of greater uncertainty than would otherwise be the case. But a certain increased cost would need to be charged to the increased social and institutional flexibility and redundancy that would be required. Aspects of these questions have been examined by Bella and Overton (1972), Bella (1974), and Watt (1974).

c) Right side. Vast unit resources permit specialization not only of individual users or harvesters, but also of vertically integrated, capital intensive industries. Large communities may be dependent on such unit resources. Whaling and cod fisheries persisted in this manner for centuries, with tuna and clupeid fisheries comparatively recent newcomers. These situations provide scope for organizational and managerial skills and it is not surprising that large corporations--both private and state operated--have evolved to dominate such systems.

Major problems arise where small enterprises, subject in some major way to a classical "market system", attempt to compete with vertically-integrated giants. The latter participate in the "planning system" (Galbraith, 1973) and can assign profits to different components of their enterprise simply by internal accounting devices. This permits the giants to interfere with and eventually cripple the small entrepreneur by indirect and well-hidden mechanisms.

Society's concerns may relate to "full utilization" in particular here. Relatively wasteful skimming of resources, disruptive pulse fishing, or overfishing to the point of collapse may well advance the welfare of the giant corporations. A more "conservative" approach might lead to appreciably greater catches and net social benefits. But the different regimes may involve quite different sets of beneficiaries. One regime might benefit mostly the distant fishing nation and the other the coastal nation.

It is clear, at the right side, that issues of sustained yield or wise management are complicated greatly by questions of who incurs costs and who receives benefits. It is with respect to such systems that the great weaknesses in the common property-free access-willing consent regimes have become well understood. International commissions and conventions have dealt--albeit ineffectively by and large--with these kinds of resources in the first instance.

d) Left side. Where unit resources are small, either absolutely or as discrete components scattered within a mosaic of other resource types, harvesting enterprises commonly spread their activities over a number of resource types. Such enterprises tend to be small, labor intensive, with a large component of husbandry and cultural practices to enhance the yield of the more desirable resources. Ultimately domestication may be involved.

Enterprises within small, mixed resource systems usually are accorded a large measure of property ownership or rights, but are subject to the forces of the market place. If that market place can be strongly influenced by large enterprises elsewhere, marketing cooperatives and marketing boards may be established to develop adequate power to balance that of the large corporations. These tend to become the large socially important issues, and a publicly supported extension system to assist in the husbandry and enhancement of small resources is usually held to be desirable.

Summarizing the right to left spectrum, maintaining near maximal catches of the preferred larger unit resource has frequently been addressed to the virtual exclusion of explicit consideration of smaller resources of the system in which the large resource dominates. This has been justifiable in the past but is becoming progressively less so with increasing demands for efficient use of all resources. On the left, unit resources as such have seldom been the primary focus of the research and information systems. Here ways have been found of managing the mix, gradually developing and enhancing the natural or semi-domesticated productive process, and assuring fair treatment in the market system.

On the right, several major enterprises may collaborate vertically or compete horizontally within the harvesting process of one unit resource. As we move to the left eventually a unit resource will not satisfy the needs of even a single small enterprise.

If the edges of this two-dimensional decision-maker's context can be characterized rather simply, what about those systems lying away from the edges? Some appropriately balanced compromise may be invoked. But there appears to be another dimension, that relates to the conceptual scope and degree of resolution of the model, and that may often be relevant to the decision-maker's needs.

6.3 Levels of Organization

The third dimension relates to a concept of hierarchical levels of understanding of natural and cultural processes. Such hierarchies have been elaborated in the sciences of economics, ecology, sociology, and also in the understanding and practice of political processes extending back at least two millenia.

A particular problem may be addressed at quite different levels of organization. A problem can be approached by focussing on the ecology of a unit resource and building the entire model around that focus. Another worker might choose to model the economic role of a group or class of resource users where the "resource" may include quite a number of ecologically different stocks or species. Or an entire fishery's role within political decision making may be taken to be important to the same problem. Suppose it were considered important to do so, how would such models be brought together explicitly?

Assuming a common level of general understanding by the various disciplines, research and information programs should show higher benefits for the research undertaken at the most relevant level of organization than for equivalent expenditures on research at lower or higher levels. This is not intended to imply that all research should be limited to one level of organization, but rather that a particular level will become the level of choice, and necessary work at higher and lower levels then be related to it.

The levels of choice appear to depend both on the expectation of yield and its uncertainty, judging from the approaches actually undertaken by different workers. The references on Figure 6.1 are located approximately according to the expectation and uncertainty associated with the systems studied by those workers. Only ecologists have been listed--a similar two-dimensional listing could be made for other disciplines, if I judge correctly.

Resource systems dominated by single, large unit resources of relatively low variability have on occasion been modelled to a useful extent by concentrating on the large dominant stocks. The populations thus studied were not very responsive to the stresses of the harvesting process, and where the latter was largely restricted to the stock under study, there the impact on the overall community was small. Ecologically such a relatively unresponsive system approximates a static system, at least when equilibrium is achieved, and thus this overall approach has here been labelled "population statics".

The term "populaton dynamics" is here used to denote systems in which strong interactions occur between populations and the interactions largely vitiate any attempts to model a single stock as though it were dominant or isolated.

Approaching a problem at the level of interacting species, using population dynamics, appears to permit profitable research of systems of higher uncertainty and of greater resource complexity than is the case with population statics. Thus if population statics is largely limited to the lower right subspace of Figure 6.1, population dynamics extends in a somewhat overlapping shell both upwards and to the left.

With larger uncertainty, the unit stocks may be characterized by low self-regulatory capabilities or perpetual response and accommodation to fluctuating large exogenous factors. Here these factors may be modelled and the variability of separte resource units related to them. Where resource systems have moderate to great complexity and numerous stocks are of interest, factors exogenous to the actual biological community may in large measure determine the characteristics and productivity of the systems. Again these whole-system variables may be modelled in the first instance. This third shell is here called "community statics" in that the overall community is not threatened by collapse, displacement or major long-term oscillation as a result of the factors responsible for the uncertainty.

Finally the fourth shell, "community dynamics", involves systems that are seldom in equilibrium, but always responding to or recovering from the ravages of major exogenous factors. Here the whole-system variables, that were relatively constant parameters in the third shell, are themselves fluctuating. Those fluctuations need to be understood, initially perhaps in actuarial context by measuring the frequency of major collapse and the length of the period of recovery, and eventually within a dynamic simulation incorporating models of causal mechanisms. The latter model will permit closer tracking of an "optimum" while the former would suffice for purposes of a government-administered insurance mechanism.

As implied in my transdisciplinary approach to the discussion of the four sides earlier, there are approaches within the disciplines of ecology, geography, oceanography and limnology, economics, engineering, sociology, political science and managerial science that can be classified according to the four shells shown in Figure 6.1. Those approaches or schools of the various disciplines that fall in the "population statics" shell, for example, should already be broadly congruent conceptually. If so, interdisciplinary research should progress rapidly once semantic differences are overcome. Conversely one should not expect to achieve significant progress quickly by convoking an interdisciplinary team that consists of an ecology expert only on a "lower right" kind of system with an economist of the "middle lower", an engineer of the "upper left", and a political scientist of the "middle right".

6.4 A Mean-Variance Context Generalized for Environmental Management

Major stresses and impacts on the natural environment may be characterized according to generalized concepts of mean and variance in a manner similar to that of fish resource yields in Figure 6.1. Figure 6.2 depicts my perception of such events as they might occur in central North America.

Again the two dimensions are (1) expectation in the statistical sense of expected average impact on the social value of the impacted ecosystems and (2) temporal variability or uncertainty that is associated with the expectation. Both horizontal and vertical dimensions extend over eight or more orders of magnitude (base 10) and are scaled logarithmically.

The variability axis may relate to both the temporal variability of the cultural loading (e.g. "pollution") process--and the background temporal variability in the ecosystem variables affected by the loading plus any interaction term. The latter may be positive (synergism) or negative (antagonism).

As with renewable resource management, developments with high impact, on the far right, will in all likelihood be addressed singly and intensively. Low impact events or processes, on the left, will often be treated theoretically and practically with other low impact phenomena. At the top the primary concern is to forestall disasters, to warn of impending trouble, or rebuild after a catastrophe. At the bottom where variability is low, the scope for the neat technological fix is maximal.

From the viewpoint of relative benefits to costs of scientific information to the decision maker (which will presumably include the informed public), developments that fall in space A will need to be assessed carefully, perhaps in an iterative ongoing process. Where the impact is on balance deleterious, careful control or detailed mitigation will be exercised at the source of the impacting process. Because of the low variability or "noise", these systems may be viewed as essentially static and quite precise optimization models may be developed to good advantage. Optimization, in any realistic sense, will be of little direct

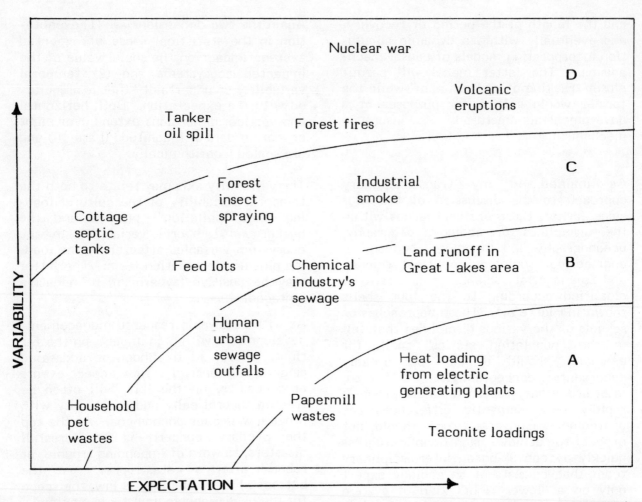

Figure 6.2 Expository model showing the approximate relative significance of expected or average effects (Expectation-E) and the temporal variability or uncertainty of those effects (Variability-V) with various kinds of pollutant loadings. E and V are variables of primary interest to practical decision-makers. Differences in the tactics employed to deal with different kinds of loadings may be rationalized in this context as shown in the text.

interest with developments that fall in spaces B to D.

With higher uncertainty and/or smaller impacts, particularly where a number of diverse initiatives impact the same system, rather more comprehensive interactive models are needed. The relative precision will not be as high as in space A partly because a measure of background uncertainty renders high precision superfluous and partly because the relative cost of comprehensive modelling increases.

With a further increase in uncertainty, as in C, the impacts likely extend over large areas or time periods, perhaps with accompanying lag effects. Here the reactive system as a whole becomes the primary focus for modelling.

Finally, as already indicated, where uncertainty is high the major interest relates to statistical forecasting and perhaps the creation of various institutional mechanisms to spread the cost of remedial action over a broad public.

I suggest that what might constitute significant environmental and ecological damage, what biological parameters would be useful, and what sampling methodologies should be selected depend very much on whether the development falls in space A, B, C, or D. To ignore such differences and propose a single comprehensive set of parameters and sampling methods is to opt for a very expensive information system with major redundancies.

6.5 Dealing with Large Scale Uncertainty

Bella and Overton (1972) emphasize that any possibility of large scale ecological irreversibility should be seriously considered. Adequate precautions should be undertaken, except perhaps in the rare case where an irreversible change would have overriding ecological advantages.

As examples of irreversible changes of sufficient concern to merit serious attention, they list:

1. Ecological components such as species or habitat types are totally eliminated.

2. A harmful condition is permitted to expand and permeate society and/or the ecosphere to the extent that it cannot be corrected.

3. The corrective action directed toward some perceived evil itself becomes a catastrophic evil--a kind of Faustian tradeoff.

4. The lag effects of good programs may years later destroy not only the good achieved but much besides.

Bella and Overton identify the essence of the environmental predicament as involving indeterminancy of the outcomes of actions that have environmental consequences. At our present state of knowledge we find it impossible to predict ecological catastrophes with any precision. Thus what is needed is a strategy that will prevent such outcomes without the requirement that each be clearly predictable. The following tactics may be applied:

1. An action restricted to part of a system is less risky than the same action applied to the entire system--spatial restriction.

2. Large scale optimization or specialization with respect to restricted goals or outputs leads to lowered resilience and lowered stability--state or organizational restriction.

3. Short-term interventions in systems are less likely to lead to irreversible outcomes than will long-term interventions--temporal restriction.

Their principle of preserved diversity states: "The chances for irreversible consequences and environmental catastrophy will be reduced by maintaining environmental diversity in two dimensions, spatial and organizational, and by elimination or constraint of those actions which have a persistent effect."

Bella and Overton's concepts provide an appropriate setting for some proposals by Regier and Henderson (1973) that were also implicitly directed toward systems of high uncertainty. These and other complementary approaches are not now receiving much attention by scientists and science funding agencies.

Recall the story of the drunk looking for a coin under the lamppost. A helpful passerby enquired--quite illogically--where the coin had been lost. Whereupon the inebriate replied that he had lost it in the park across the street but it was too dark to look for it there.

Similarly much of the current ecological attention is directed to the simple, more easily understood issues where the major scientific challenge no longer exists. But better to work in the light even with low probability of additional payoff, than to grope in the half light of uncertainty for badly needed insight .

6.6 Relating the Mean-Variance Space to Planning Concepts

Some planners recognize a number of levels of decision making according to the relative degree of scientific objectivity vis-a-vis willed purposiveness in society's regulatory process. Four levels may be specified, for example by Jantsch (1972), as follows: empiric, pragmatic, normative, and purposive. (See also Chapter 10). These levels may be perceived as mapping into different levels of the conventional hierarchic political decision-making process.

Issues that can be adequately treated empirically are usually dealt with a at low level of the power hierarchy, always provided that adequate policies, norms and technical competence are already in place. It happens that resources or impacts at the bottom right hand corner of the E by V space, Figures 6.1 and 6.2, tend to meet these preconditions for empirical management more nearly than any other of the E by V subspaces. As one moves to the left or upwards in that space, adequate policies may not be in place, norms or standards may not have been specified, and the scientific-technical competence may be weak due to the immaturity of the relevant science. Any and all of these shortcomings tend to ensure that such issues will be referred upwards in the decision-making hierarchy. At the top right hand corner, where high stakes and high risks are involved, a large element of purposiveness at high levels of the technical-political hierarchy may be invoked. Such decisions may emerge as decrees without a great deal of rationalization attached to them. Table 6.1 is an attempt to explain some further considerations involved in these matters.

It may be granted that the concepts depicted in Figures 6.1 and 6.2 have some broad heuristic relevance, but it may be queried whether they have any direct practical application. I suggest that they do, in a number of ways, two of which are sketched below.

Minimally an attempt by workers, addressing the same event or phenomenon, to locate it in such E by V space could serve as a "perception check". R.T. Oglesby has pointed out that large thermonuclear electric generating installations might be placed in the lower right hand corner by persons who are not much worried about the risk of catastrophic accident while persons who are more concerned about such risks might locate it near the top right hand corner. Consistent with the different perceptions, the trusting, melioristic persons would urge safeguards of an

Table 6.1 THE RELATIONSHIP OF E x V CLASSES OF FIGURE 6.2 TO JANTSCH'S CONCEPTUAL LEVELS SHOWN IN FIGURES 5.1 AND 5.3 IS ILLUSTRATED AS A "TRANSFORMATION."

E x V classes	Stresses-- ecosystem interactions	Geographic scope of loading	Temporal aspects of loading/control	Institutions: Jantsch's levels
				Empiric
A	major discrete impact	often small, or point source	real time	
B	several linked stress interactions	diffused or distributed moderately	annual regimes, important	Pragmatic
C	stress(es) interacting with stable system	spatially uneven, extensive	system lags to be predicted	Normative
D	interactive system unstable	intensity and area indeterminate	early warning of impending disaster	Purposive

empiric sort while the fearful, pejoristic persons would urge high level policies-- perhaps to ban such installations altogether.

Provided that there are no major perceptual ambiguities, the E by V space as structured in Figures 6.1 and 6.2 might serve as a screen to identify inter-disciplinary concepts and intradisciplinary approaches that would seem best suited to address the particular issue. To complicate--such a first-cut screening could help to identify rapidly a first choice as well as a second choice, and both might be pursued concurrently.

6.7 Map, Monitor, Model

The science particularly relevant to the understanding and management of any system of renewable resources and the natural environment may be perceived to consist of a dynamically evolving set of three activities: mapping, monitoring and modelling. (The terms surveying, surveillance, and assessment, implying a conceptual framework, have similar connotations. See Regier and McCracken 1975.) Each member of the set has a useful function--to neglect one in order to favor another leads to inefficiences from a broad practical viewpoint in the sense that benefits (information and insights) may dip below costs (efforts expended in achieving the information and insights). Yet certain types of practical questions can best be answered in the short run by one or other components.

Questions of property ownership or stewardship responsibilities defined spatially--a kind of territoriality--call for development of appropriate *maps*. In addition, the valuation and utilization of a resource or assessment, and regulation of an environmental impact by some other use of the resources or space, require appropriate identification of amount or concentrations and their distribution.

Assuming some spatial mosaic of rights or interests sanctioned by law or custom, new enterprises or innovations in the use of commonly-shared assets will influence other interests. Time trends of important variables known to be related to particular activities, *monitored* at selected points, provide a basis by which vested interests may assess whether they are benefiting or suffering as a result of their own and others' actions. Regulatory protocols may be established that require appropriate corrective action by the offending parties when indices based on monitored data reach predetermined levels (e.g. maximum sustained yield, tolerance limit, carrying capacity, etc.).

Ultimately neither detailed maps nor balanced data from monitoring (or surveillance) will suffice for practical decision making. Use of shared aspects of the environment or resource may become so complex that the overall effect may include major interactions of the separate effects. Cause-effect modelling, involving identification of hypotheses, experimentation in field and laboratory and ultimately dynamic simulation with

appropriate testing, may be necessary. This may be particularly true if litigation or legislative action are involved.

Even at an early primitive stage of the development of the science of renewable resources and natural environment, a simple conceptual model of local causality does underlie the selection of qualities to be mapped and quantities to be monitored. Data from the latter two sources almost inevitably contribute to better understanding even if modelling and experimentation are not formally undertaken.

Eventually with increased sophistication some division of labor may be desirable with cartographers and sampling specialists assisting with the mapping, statisticians and time series experts with the monitoring, and laboratory and field experimentalists, systems and computer specialists helping with the modelling. In the industrially developed parts of the world such a niche differentation involving some "interactive segregation" is now occurring. But it would of course be counterproductive to induce intense specialization.

In general the problems of sampling techniques, sample survey design, data management, statistical estimates, and reporting mode differ in their details between mapping, monitoring, and modelling. The three components relate to different aspects of the overall needs of practical people. It would usually be relatively costly to devise a single homogeneous program to obtain data concurrently for all three roles. It would also be costly to separate the roles entirely. Efficiency requires dynamic collaboration within the evolving system with each component containing suitably specialized expertise and retaining some measure of independence (Figure 6.3).

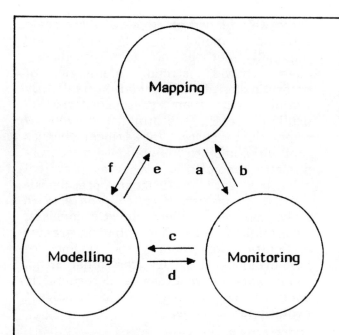

Figure 6.3 Interactions between three major components common to the science and practice related to renewable resources and the natural environment.

a: A map shows the spatial distribution of some quality at a specified time or more usually the average distribution within a specified period. If mapping is undertaken repeatedly, trends through time can be inferred and thus the mapping approach converges on the usual role of monitoring. Generally, mapping is repeated at longer intervals of time to produce regional baselines; monitoring a limited set of points then permits useful and economical interpolation and some extrapolation.

b: Monitoring provides data on trends of selected variables. If law or custom allows these to be influenced to move far from their baselines, particularly if the trend is consistent, new maps will need to be produced. Thus monitor data can help to indicate when existing maps are obsolete.

c: From the viewpoint of understanding a system and its responses, particularly if it possesses strong homeostatic mechanisms, data from an extensive monitoring network can indicate some of the time characteristics (e.g. cycles, lags) of the system. Also sets of variables may be analyzed statistically to yield hypotheses concerning causal interactions.

d: A fuller analytical understanding of the system permits a more incisive and efficient selection, deductively, of variables and indices for monitoring purposes.

e: As with "d" above, a well-tested model may be analyzed or exercised to yield information concerning which qualities have high information content for mapping purposes.

f: Analogous to "c" above, the approximate spatial limits of the system of interest may be inferred from a map plus some previous insight of the general ecology involved.

The three scientific components-- mapping, monitoring, modelling--are all essential for resource and environmental systems regardless of discipline.

6.8 New Initiatives Needed

In the hierarchical systems concept sketched by Jantsch (1972), and with special reference to cultural impacts on aquatic ecosystems, we now are reasonably competent at the scientific-empiric and applied-pragmatic levels with respect to ecological, economic and psychosocial sciences. Unfortunately there is little effective communication with workers expert at the normative and purposive levels of decision making. It is these higher levels of the hierarchy that should assist in clarifying what kind of ecological information really is needed. This demand-pull approach may be contrasted with the supply-push method of most applied ecologists. Many cannot make a clear case for the practical relevance of their insights and information but offer up their hard-won nuggets in the hope that someone somewhere will put them to good use.

Of course, a strong case can be made for some appreciable support for curiosity-directed scientific research. That is another matter. If our concern here is to develop more effective means of forestalling or mitigating deleterious cultural impacts on the environment, a demand-pull form of mobilization should now be quite effective.

When I plead for more funding for study of the consequences of high environmental uncertainty or rapid irregularities of major natural factors influencing fisheries and their environment, I do not intend to imply commitment to an approach that geographers designate as "environmental determinism" (Burton, Kates, and Kirkby 1974). Though humans may be strongly influenced by environmental factors, singly and collectively, and must accommodate our actions to many of them, this does not imply to me that climate *controls* my life, e.g. in the sense of Winkless and Browning (1975). There is a great difference between *affecting* and *controlling*.

Some well-funded attempts at large scale dynamic modelling globally and regionally--perhaps all such attempts-- have shared strong elements of determinism. This isn't bad, in itself. By working out their presuppositions in explicit detail, determinists even if scientifically correct may trigger changes that will cumulate to invalidate the deterministic predictions. Whereas practical scientists with evolutionary presuppositions (e.g. Jantsch 1975) may be able to make good use of deterministic analyses, determinists may have much greater difficulty understanding the findings of the evolutionists. On the other hand, some workers who use a deterministic approach do so as a tactic to effect evolutionary change.

Concerning matters of high uncertainty, especially those that fall toward the upper right edge of Figure 6.1, it may be particularly difficult to judge whether a particular approach is soundly scientific or not. Perceptions, cognitions, and unstated presuppositions may play particularly important roles in the approach taken. The larger purpose of the worker in undertaking the study may in part determine the methodology and data collected. On such issues it may be difficult to discriminate between science and science fiction. Granting these and other scientific difficulties, nevertheless it is timely to direct more attention to fisheries issues of higher uncertainty and major irregularities.

7

life histories of typical fisheries

No intensive fishery on a major resource has come into being overnight. Some have developed gradually--though each with some historic irregularities in the development process--over a period of centuries (e.g., the North Sea, the Bodensee and the Mediterranean). Others were fished at low levels of intensity for long periods of time and then subjected to fairly rapid increases in intensity within a few decades (e.g., the Northwest Atlantic and Northern Pacific). More recently, roughly comparable increases in levels of fishing intensity that occurred over centuries in some ecosystems have been achieved in about a decade (e.g., off Peru and Northwest Africa). If there were any major shelf-seas fishery resource system left fully accessible to the world's fishing fleet, a fishing-up operation could be mounted to achieve the system's maximum sustainable yield within a few years. Aside from the rate at which such evolution occurs, has the process been more or less similar with different fisheries?

7.1 A Model of the Evolution of Fisheries

Larkin and Wilimovsky (1973) have sketched an admittedly idealized version of the evolution of a fishery, as follows:

In the initial stages, a fishery will usually be centred on relatively abundant, ecologically dominant species. Characteristically, these support substantial rates of harvest, and may even be sufficiently resilient to regain high levels of abundance following relaxation of fishing intensity by which they had been depressed to low levels for long periods. As species, therefore, they are in little danger of extinction; but in the development of a fishery some may be depressed to levels that do not produce the greatest sustained annual harvest.

Much of the effect of a fishery on a single species depends on the range of the fishermen in relation to the range and rates of dispersal of the fish. A highly

localized fishery, based on a widely distributed stock, has no long-term effects on
stock abundance, but a wide-ranging
fishery may greatly reduce many local
stocks. The impact of fishing is thus a
consequence of the interaction of the
spatial distributions and mobilities of the
fish and the fishermen. For these reasons
it is characteristic of the early stages of
development of a fishery that there is an
almost linear relation between effort and
catch. As catches decline near the fishing
ports, the fleet fishes further away, trad-
ing off travelling time against greater
catches from unharvested segments of the
stock.

As the yield of preferred species declines,
the fishermen may turn in part to other
species that are taken jointly and are
acceptable to consumers. It is characteristic of fisheries for almost any species
that the current estimate of a maximum
sustained yield will be exceeded; but in
fisheries for most areas the total yield
will gradually increase. This is accomplished by progressive harvesting of more
species. The harvest increasingly constitutes a less selective removal, with less
emphasis on taking particular species at
particular sizes. The rate at which this
process develops largely depends on the
demand for fish and the broadness of the
tastes of the consumers. Ultimately, at
least in theory, the rate of harvest is
dictated by economics. As the advanced
stages of harvesting are approached,
theory suggests that fluctuations in
abundance of a particular species become
more and more extreme as a prelude to
their elimination. Sudden rearrangements
of relative abundance are to be expected.
At present we have only a weak
perception of the effects of heavy
harvesting on the structure and stability
of aquatic communities.

There are two major qualifications to be
attached to the foregoing as a summary of
how unmanaged fisheries affect the
species harvested. Firstly ... (their)
impact on stocks of fish is usually
obscured in some degree by natural
changes in the physical environment. For
example, a fishery may be initiated when
it is noticed that there is an abundance of
a particular species, but such abundance
may be a consequence of unusual and
short-lived oceanographic conditions.
Similarly, a fishery of long standing may
collapse in a few years as a result of
oceanographic changes. At all stages in
the development of a fishery, fluctuations
in environmental conditions may augment
or offset the impact of harvesting on
abundance. When a fishery is intense and
suffers environmental change, it is
expected that there will be wide, and
sometimes almost irreversible, changes in
the aquatic community structure.

Secondly, ... most of our appreciation of
the impact of fisheries is based on
experience in the north temperate zone,
where there are strong seasonal elements
and relatively small numbers of abundant
species. Tropical environments are
considerably less well known, and their
fisheries may be characterized by
relatively large numbers of species, none
superabundant. In these circumstances,
our temperate zone concepts should
perhaps be superseded by an approach that
treats organisms by communities and
trophic levels rather than by species.

The whole process of interaction between
a fishery and a group of harvested species
may be visualized as a system that
evolves with mutual responses, always
with a background of environmental
variability that influences the biological
side of the system, and a context of social
and economic factors that influence the
human side. The general properties of
such systems are not well known, but they
seem to trend in the direction of greatest

productive efficiency if left to their own devices.

Leaving the last sentence aside for the moment (see section 7.3) that account is both concise and quite comprehensive as it concerns an ecological viewpoint of the overall evolution of a fishery. It agrees well with a more limited and detailed analyses of the evolution of fisheries on salmonid communities in oligotrophic lakes by Loftus and Regier (1972) and Regier (1973). McHugh (1970; 1972) has published broadly similar observations and inferences.

7.2 Standard Advice on Scientific Support Services

Another approach to understanding how fisheries workers have perceived the evolution of fisheries is to consider what may be implied by the kinds of advice directed toward the development of useful scientific support services.

Lucas (1973) advised as follows:

For fisheries in their very early stages, and for future fisheries, it is essential to have as soon as possible some idea of the potential, so that it is not rashly exceeded. To be timely, these may need to be obtained as relatively crude estimates by means of exploratory fish surveys, egg and larval surveys, and even primary productivity estimates.

For more mature fisheries, and more precise estimates of maximum sustainable yield, classical data are required concerning catches, size compositions, effort applied, growth rates, distributions and movements, etc., together with ecosystem studies relevant to food and predators.

While classical models are available for estimating maximum sustainable yield, there is need for further research into the most suitable models for particular circumstances, and especially into the circumstances in which intensity of fishing affects not only the stock but also recruitment.

Environmental effects naturally vary greatly, in time and space, for good and bad, while man's activities through pollution risk exerting a unilateral and undesirable effect on fish, other organisms and man himself. Much more research at the highest level and, in due course, monitoring are needed if environmental effects are to be incorporated in the models for prediction, and release of pollution regulated at appropriate levels.

The above advice again has a strong biological-ecological bias, perhaps justifiably since the title refers to "scientific advice" and many workers in the natural sciences do not include "social science" with "science".

Also the quote seems to relate most clearly, though implicitly, to temperate, shelf-seas situations and as such is generally consistent with the relevant parts of Larkin and Wilimovsky (1973).

Turning to a report of an ACMRR working party of which I was a member (Lucas et al. 1974), the topic of "the qualities and deficiences of scientific advice" was addressed under the subheadings: timeliness, accuracy and precision, clarity and ease of comprehenion, scope and relevance, acceptability and credibility, with a further comment on communications.

The sequence was presumably not intended as specifying the order of importance, from highest importance at the beginning to least importance at the end of the list. Nevertheless, to read the list in reverse order (except that timeliness should be at the beginning of the new order as well) would correspond more closely with the approach likely to be employed in the "transfer sciences" now emerging (see Chapter 5). By contrast, marine fishery workers have generally approached these matters in the sequence given by the ACMRR report.

7.3   More Complications

Recall that Larkin and Wilimovsky (1973) stated:

The general properties of such systems are not well known, but they seem to trend in the direction of greatest productive efficiency if left to their own devices.

If "greatest productive efficiency" is to be interpreted with respect to local economic criteria, this statement would clearly be at odds with the current, widely shared view that economic inefficiency follows inexorably in the wake of a thorough-going laissez-faire approach within a regime characterized by common property, open access and willing consent.

On the other hand, if "productive efficiency" is to be interpreted ecologically, then the authors appear to have ventured an educated guess. I haven't seen any data that support the concept that a laissez-faire fishery would self-regulate approximately at the level of maximum efficiency in production of catchable fish flesh.

Dickie (1970) addressed a related question:

In principle it is already possible to calculate the minimum food intake to support a specified amount of production at various levels of food chains. However, to estimate potential fisheries yield (with useful production efficiency) requires that this (reasoning) process be reversed and that the production or that part of it realized as yield be predicted from an observed food abundance. This latter process raises problems yet to be solved . . .

Perhaps Larkin and Wilimovsky were appealing implicitly to the simple trophodynamic model according to which one should be able to harvest the higher levels of a trophic pyramid and take their place as top predators. Since a fish predator devours 5 to 10 times as much flesh as it assimilates for body growth, this kind of "fishing-down" the trophic pyramid should yield progressively greater potential yields. I have been looking for evidence in support of this theory for about ten years but have yet to find it. With respect to growing protein needs, it would certainly be reassuring if it were true or if it could be realized on a grand scale by some inexpensive ecological fix. As it now stands, it is a troublesome myth.

Another major complication overlooked at that time by Larkin and Wilimovsky (1973), as well as by Lucas (1973), relates to the need for rehabilitative ecological measures for some fisheries following relaxation of major, longterm stresses. In their absence, effective recovery may lag for decades after the definitive political decision is taken.

7.4   Information Related to Fishery's Life History Stage

For the purposes of this section consider a fishery and fish resource that is rather large and well-behaved (other possibilities are examined in Section 7.5). Here fish resource systems like those of temperate shelf seas and some of the world's great lakes might serve as examples. The harvestable surplus of ecological production is not highly responsive to the stress of fishing at light to moderate intensity. Abundant stocks of many of the larger species may accommodate considerable fishery exploitation without serious risk of collapse, though eventually any fish stock will collapse or fluctuate violently given excessive exploitation. These presuppositions in effect imply that the fishes' natural environment is neither highly "unstable" (in the sense of large-scale, unpredictable fluctuations), nor highly "stable" (in the sense that environmental fluctuations almost never exceed a small scale). The fishery does not redirect its preferences nor alter its basic methods rapidly and haphazardly in time and space. These idealized characteristics are not very different from the ecological circumstances and exploitive history of the temperate shelf seas and temperate large lakes of the world prior to about 1950.

Table 7.1 lists six rather idealized stages in the maturation of such a fishery. In the second column, I have specified the kind of ecological information that would appear to be most directly relevant corresponding to each of the six stages.

The term "most profitable stocks", used in Table 7.1, should be explained. Simply put, it is the maximum difference between what the fishing enterprise receives, for its landed catch of particular stocks, less the corresponding costs of fishing for them. Thus fishing near shore for less preferred abundant stocks with inexpensive vessels and gear may be as profitable as fishing far off shore for more highly preferred stocks with large vessels and costly gear. Use of the concept "most profitable stocks" thus permits a broad generalization of the idealized sequence over both inshore and offshore stocks whether or not they historically progressed through the six stages temporally in phase.

Stages 1 to 6 in the column one of Table 7.1 are not to be considered discrete in the sense that activities listed in each begin and end in that particular stage. Rather the stages identify where in the sequence a particular activity is perceived as being most creative, innovative, and potentially useful. As the concepts and methods become well-understood, they are transferred into practice and are thereafter accepted as part of the shared common background of the scientific professionals.

Table 7.1 focuses on relevant biological information and makes no mention of other kinds of information (economic, technological, social) also useful to fisheries managers at the various stages. The relevant kinds of limnological and oceanographic information could easily be sketched by fishery ecologists. Corresponding sequences can also be sketched for economics, fishery technology, and policy sciences that deal with regulatory processes.

Most of the large fisheries of the world are now past Stage 3. Unfortunately many fishery biologists are lagging at least one stage behind that which would be most directly relevant according to Table 7.1. Very few are one stage ahead-- where innovative science should be. But more of this later.

7.5 Information Related to Levels of Environmental Stress

Many aquatic ecosystems that have yielded fisheries resources have also

Table 7.1 IDEALIZED SEQUENCE OF STATES IN FISHERY DEVELOPMENT OR "MATURATION" PROCESSES, WITH IDENTIFICATION OF THE KIND OF BIOLOGICAL SCIENCE MOST DIRECTLY RELEVANT TO EACH STAGE.

State of fishery resource use		Relevant, innovative biological information
1.	Exploration, trial fishing, adaptation of gear	. . . Linnean taxonomy, geographic and seasonal distribution of larger individuals of profitable stocks.
2.	Initiate and expand systematic exploitation of *most profitable stocks (MPSs)*	. . . (a) Identify life history stages of MPSs, and their spatial and temporal distributions, especially the critical reproductive stages; (b) First rough assessment of potential yield of MPSs.
3.	Intensify fishing on MPSs to full utilization; expand to *less profitable stocks (LPSs)*	. . . (a) Populations dynamics of MPSs to yield separate estimates of optimum or maximum sustainable yield; (b) Identify and clarify strong ecological interactions between species (MPSs and LPSs) in order to adjust separate yield estimates on an ad hoc basis.
4.	Full and intense exploitation of all marketable stocks	. . . Dynamics of the fish taxocene and/or ecosystem to investigate overall profitability of yield of alternative management options.
5.	Chronic overfishing, eventually recognized as such	. . . From whole-system theory based in part on study of replicate systems, estimate the likelihood of major irreversible consequences, such as species extinction.
6.	Resource rehabilitation	. . . "Ecological engineering" to correct major ecological deficiencies and deformities and/or "therapeutic ecology" to aid natural recovery processes.

served many other purposes, especially in industrialized parts of the world. The life histories of some important fisheries have been strongly influenced by such stresses.

Again consider a somewhat idealized case involving the effect of a growing human settlement, with surrounding geographic "resource shed", on a lake or estuary nearby. As before, the aquatic ecosystem is taken to be fairly unresponsive and can withstand moderate stress without major disruption. Assume that no very intensive fishing is permitted. Instead the aquatic ecosystem is subjected to quite intensive sewage loading, shoreline restructuring, bottom dredging, loading by atmospheric particulates from an industrialized society, heat loading, and biological loading with exotic species. A sequence of stages roughly comparable to those of Table 7.1 is shown in Table 7.2.

Table 7.2 IDEALIZED SEQUENCE OF STATES IN THE USE OF THE AQUATIC ENVIRONMENT AS A SOURCE FOR ABIOTIC RESOURCES, AS A SINK FOR WASTES, ETC., WITH IDENTIFICATION OF THE KIND OF SCIENCE PARTICULARLY RELEVANT TO EACH STATE.

State of aquatic ecosystem abuse		Relevant, innovative biological information
1.	Build harbors and canals, drain wastebearing surface waters, withdraw water for local distribution	Linnean taxonomy to identify which of the most profitable fish stocks (MPSs), if any, are seriously affected by determining whether their spatial niches encompass the small areas affected.
2.	Build dams and levees, dredge harbors, construct piped sewage and water systems, widespread swimming and boating	(a) Identify life history stages of MPSs with their spatial and temporal distributions, particularly of the critical reproductive stages, relative to heavily stressed habitats; (b) First rough assessment of overall impact on several of the MPSs.
3.	Restructure shorelines, drain marshes, treat sewage and drinking water chemically, construct and tend beaches and yacht basins, loading by atmospheric particulates that fall into the habitat	(a) Approximate assessment of the impacts of separate anthropogenic stresses on the MPSs; (b) Joint assessment of the two or three most destructive stresses to determine the general nature of any ecological interactions among them, to judge whether ecologically-oriented regulation is needed.
4.	Widespread and intense multiple use, with some effort made to separate users spatially and/or temporally	More precise assessment of separate and joint effects of stresses to consider various alternative use and management options, which requires a fairly comprehensive understanding of ecosystem structures and processes.
5.	Chronic overuse with major undesirable interactions between various users' ecological impacts	On the basis of whole-system theory, predict the demise of one valued ecosystem component after another.
6.	Habitat rehabilitation	Ecological engineering to correct deficiencies in ecological structures and/or therapeutic ecology to aid natural recovery processes.

A comparison of Tables 7.1 and 7.2 shows that I have defined the use stages in both cases so that the biological information stages within the two sequences resemble each other. If it happens that fishery use and environmental abuse proceed historically in phase through Stages 1 to 6, then the kind of biological (and other) science most relevant at each stage is conceptually similar for both the use and abuse cases. Where they proceed out of phase, it might at first glance appear that the science relevant to the particular stress process which is lagging can afford also to lag behind the other. But this overlooks the fact that a species' habitat and functional niches are closely interrelated and both are influenced by each of the major types of man-caused stress. Hence whichever stress process leads in the "maturation sequence" determines the type of biological perspective most effective for gaining an understanding of the ecological response to either or both stress types.

7.6 Information Related to Ecological Properties of the Resource

It may now be apparent that Chapter 7 has dealt primarily with the temporal case history aspects of the subject matter of Chapter 6.

For present purposes consider the two dimensional expectation by variability space to be separated into four quadrants: (a) large and well-behaved, (b) small and well-behaved, (c) small and poorly behaved, and (d) large and poorly behaved. An ecosystem and/or resource harvesting system that is "well-behaved" has high self-regulatory capabilities, rapid recovery, and high resilience.

a) The idealized case history sequence of sections 7.1, 7.2, and 7.4 seem to relate most directly to the large and well-behaved type. Because such individual resources have major practical importance, it may be inevitable that the sequence of practical use and relevant ecology shown in Table 7.1 will be traversed at least to Stage 4. If the resource becomes seriously degraded through overfishing, or some other kind of over-intensive use (see Table 7.2), Stages 5 and 6 may be realized.

b) With small, well-behaved resources each individual resource may be too small to justify intensive individual study. There are "economies of scale" in ecological science, hence a research approach that is economically practical for a large resource (e.g. stock-by-stock population dynamics) is excessively costly relative to benefits with a small resource. For these and other reasons, workers in small resource and environmental systems have tended to skip through steps 1 through 3 (Tables 7.1 and 7.2) rather rapidly and to settle down to relevant work in steps 4 to 6. Some important scientific beginnings to Stage 6 may be found scattered throughout the north-temperate zone. With respect to very small systems, such as Chinese aquacultural ponds, a well-developed methodology has existed for centuries, though it has apparently never been well described scientifically.

c) Small poorly-behaved resources include minor flood plains, shallow lakes of high latitudes, and some waters intensely used or abused by man. Whether the cause of the high uncertainty is natural or technological, the most relevant biology appears to be that of Stage 5 or 6 (Tables 7.1 and 7.2). To address the issues using only the biology most relevant to Stages 1 to 4, as they applied in type (a), would involve relatively large research costs.

d) Large poorly-behaved resources must be dealt with in the first instance by policy and management sciences; ecological and other sciences would be used to provide some forecast data for consideration in the context of the policy and management sciences. An expanded Stage 5 seems most relevant; Stage 6 is largely ruled out where the forces causing the uncertainty are beyond local control.

8
toward balanced scientific services

A number of perspectives and concepts have been presented in previous chapters; they were usually related to current scientific and practical reality in general terms. Here some of them are again taken up, perhaps in altered guise, in order to relate them more clearly to major problems with renewable resources and the natural environment--particularly with respect to fisheries.

In the next chapter are sketched the scientific challenges for the resource and environment aspects of Canadian fisheries, as they were perceived by research planners who shared in some general way the perspectives of this book. In the final chapter some implications of the current evolutionary transformation for institutions, particularly universities, are proposed.

8.1 Encouragement of Creativity

Nowadays science in general is not enjoying a "good press", at least not in the more sophisticated press. Few non-scientists would now feel any great urge to jump up and cheer in order to rally scientists to high levels of creativity.

Many scientists themselves would not seek support for, nor even condone, a "creativity" that implies totally unfettered scientific freedom to do anything with public funding. Under what circumstances or which criteria do we feel that creativity should be encouraged?

Hetman (1973) addressed problems related to this matter. Several excerpts follow:

Modern science grew up as a combination of philosophy and technique, giving rise to a concept of knowledge characteristic of itself. The philosophical aspect of science has had various meanings over the centuries; the most significant is still that derived from the Enlightenment which considers science as a liberation of men from dogma and superstition. The second aspect is knowledge considered as a good in itself; this ideal serves as the main ideology of university-based science. The third aspect is technique, which appears as a justification of science by its contribution to industry and economy. Though independent in principle, these three conceptions of science are not necessarily inconsistent. Hence, it has been possible for scientists sincerely to claim a bit of all these; technique for external support, knowledge for professional purposes and philosophy for idealism. The recent ideological problems of science have arisen because of the very success of this mixture. The quantitative expansion of science since World War II, the increase in size and scale of capital investment for individual R and D projects, the change from " academic science" to the "techno-scientific complex" have transformed the socio-economic character of science. (Hetman, 1973 pp. 26-27).

Concern over frustration resulting from technology is voiced on various grounds. Those most often repeated hint at:

a) the accumulation of means of warfare and in particular the menace of nuclear destruction;

b) the disruption of ecological and environmental equilibrium through polluting and detrimental technologies;

c) the depletion of natural resources through the unheeding exploitation of nature;

d) the surrender of scientists and technicians to a political system and its needs;

e) the discrepancy between technological change and the ethical progress of man;

f) the search for power on the part of technology people.

More sobering appreciations tend to take into account past achievement of technology and express the need for a thorough assessment of the positive and negative effects of each technology, with a view to putting certain limits to technological developments in the future.

Many scientists and technologists--but also politicians and economists--favor the idea that more and not less scientific research and technology will be necessary to reduce the negative side-effects of existing technologies (Hetman, 1973, p 33).

Thoughtful, articulate people are becoming increasingly critical of the "techno-scientific complex". Environmentalists who have become politically active have often found other politically active scientists facing them across the barricades. These differences are coming to be sufficiently great that united support among scientists for undifferentiated "scientific creativity" may no longer be possible or desirable. It is as a protagonist for an environmental ethic related to the needs of a freely evolving, heterogenistic society that I can seek--with a clear conscience--that society's support for creative scientific research on related matters. It is within this context that I accept the model by Gregory (1975) (see Chapter 5), which recasts the "pure-versus-applied" controversy into a trichotomy: (1) analysis-based or curiosity-motivated; (2) transfer or mission-oriented; (3) action-based or program-dictated workers (see also Regier 1974; Regier and McCracken 1975). The transfer group mediates between the two other groups, but also interacts more fully than the other two in interdisciplinary and/or transdisciplinary work.

8.2 Structure and Process
or
Form and Function

About 150 years ago--with the rise of rational positivism in reaction to dogmatic theology and romantic philosophy--Comte and others sought to reduce *to simple formulas all phenomena even those of the most complex character . . . He believed, for instance, that every being, and especially every living being, can be studied from two sides, the static and the dynamic--that is to say, as potentially active and as actually active. Thus, biology has a static side, anatomy, and a dynamic side, physiology, and other sciences in like manner . . According to Comte, all science should be classified after the method employed by botanists and zoologists; by this method we get six separate branches of science: mathematics, astronomy, physics, chemistry, biology and . . . sociology. Each of these sciences is based on all the previous ones in the series and cannot be mastered without a knowledge of them. (Nordenskiöld 1928, p. 444).*

If we were to interpret "astronomy" in that sequence as including geology and cosmology, then this must all sound like the conventional wisdom of the 1970's for many biologists. Nevertheless there has been a decided swing away from simplistic positivism in all the sciences. To what extent, then, are such dichotomies as form and function, and scientific hierarchies likely to become constraining? Clearly the approach is well suited to analytical reductionism--is it likely to interfere seriously with an attempt to view reality more holistically, as in a mutual causal paradigm (Maruyama 1975, see Chapter 2). A partial compromise, that does not escape fully the reductionist trap, is to treat such dichotomies or proposed hierarchies as alternatives within a broader dialectical approach. It is in this way that I seek to justify using the structure-process dichotomy in Chapter 4 and more implicitly throughout the book.

In a period of rapid social and scientific transformation an overemphasis on formal conventional structure of any kind is clearly reactionary. Also a premature attempt to structure what is emerging may interfere with the process before it is near completion. The strong current emphasis on process throughout western society is an indication that the larger process of change is not nearly complete (Pirsig 1974; Rotstein 1976).

The characterization of "transfer sciences" by Gregory (1975, see Chapter 5) implied to me a kind of transdisciplinary quasi-discipline or temporary macrodiscipline that is created for a major purpose and eventually dissolves when that purpose is achieved or loses visibility for whatever reason. Like institutions, disciplines tend to become structure-oriented and thus persist, beyond their time, as well-endowed monuments to yesterday's problems. The more process-oriented approach is to be satisfied with pragmatic, temporary quasi-disciplines. Like scientists, disciplines are born, become mature, may produce something of lasting value, then gradually phase into retirement, and are eventually interred with full honors to live on in historical studies. Perhaps one of the processes the biologists understand well--the individual human life--could serve more broadly as a model of our scientific institutions and disciplines.

8.3 Functional Hierarchic Frameworks

A framework obviously relates to form and structure, but the modifier "functional" again implies a compromise. The realistic option now seems to be to use hierarchic concepts despite their deficiencies and to take special care not to become entangled in them. In the first instance, a scientist who wishes to be free of such constraints should wander about in the different levels perceived as being hierarchically arranged and simply ignore

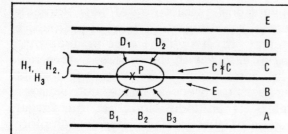

Figure 8.1   Sketch of tactical alternatives in the search for hypotheses and models within a levels-of-organization paradigm.

a: Hypotheses H_1, H_2, H_3 . . . related to level C already exist in the literature, in one form or another, and may be interpreted or reformulated with respect to P.

b: Inferences B_1, B_2, B_3 . . . at one level of organization below C may be agglomerated to form an hypothesis for P. This type of bottom-upward modelling has been much in evidence in ecology

recently. The efficacy of this approach is under continual debate in science; in general it seems a very expensive way to proceed.

c: Existing hypotheses or models at level D may be analyzed or disaggregated to provide C-level hypotheses deductively. This approach is underexploited in fisheries science.

d: C and C refers to the classical compare-and-contrast approach which should be applied more widely in fisheries work.

e: E denotes an empirical, experimental, or trial-and-error way of proceeding. This is the basis of much that now exists as fisheries science, though it remains loosely formulated.

f: Finally, there remains the intuitive blinding flash that some workers, subject to strike-it-rich hopes, expect to happen.

any "Keep Out" signs. The need to become sufficiently competent in some aspects of science and to do productive, disciplined work need not imply acceptance of pro forma boundaries to one's perceptions and perspectives.

Consider Figure 8.1, in which the framework is the usual hierarchy within an ecological system of perceptions: organisms, populations or guilds, communities or ecosystems, biomes, etc. (My point can be made with respect to any one discipline, e.g. ecology, or with respect to an interdisciplinary framework; but it may be more easily recognized if made with respect to the one discipline.)

A liberated ecologist is confronted with a particular problem, P. In a preliminary analysis (see Chapter 6), he decides that it

can be addressed most directly and effectively using science (inferences and methods) at level C. This might be the ecosystem level. In the case where the science is not yet ready-made, the worker must search out useful hypotheses or develop some of his own, for testing and eventual application. The tactical alternatives are suggested in Figure 8.1.

a) Hypotheses H_1, H_2, H_3, . . . related to level C already exist in the literature, in one form or another, and may be interpreted or reformulated with respect to P.

b) Inferences B_1, B_2, B_3, . . . at one level of organization below C may be agglomerated to form an hypothesis for P. This type of bottom-upward modelling has been much in evidence in ecology recently. The efficacy of this approach is under continual debate in science; in general it seems a very expensive and ineflective way to proceed.

c) Existing hypotheses or models at level D may be analyzed or disaggregated to provide C-level hypotheses deductively. This approach is underexploited in fisheries science.

d) C and C refers to the classical compare-and-contrast approach which should be applied more widely in fisheries work.

e) E denotes an empirical, experimental, or trial-and-error way of proceeding. This is the basis of much that now exists as fisheries science, though it remains loosely formulated.

f) Finally, there remains the intuitive blinding flash that some workers, subject to strike-it-rich hopes, expect to happen.

8.4 Trends in Quantitative Sophistication

How the needs for particular kinds of scientific information are related to stages of development of resource systems, and/or state-of-stress loading on the relevant ecosystem, was addressed in Chapters 6 and 7. How quantification proceeded in a particular tradition widely applied by fisheries workers, i.e. population dynamics, was illustrated in Chapters 3 and 4. Now the focus shifts more directly onto quantitative sophistication as such, and is justified if for no reason other than that many ecological scientists are less interested in the concepts of ecology than in the quantitative methods used in the science.

Figure 8.2 and the following text is based on the work of an interdisciplinary team and the consensus text was proposed as broadly relevant to most if not all areas of scientific and technical studies related to the natural environment and renewable resources (see Regier et al. 1973 and Regier et al. 1974).

In general quantitative aspects of ecology have developed through time approximately as shown in Figure 8.2. Perhaps because the overall trend was largely uncharted and unplanned, the process has been uneven among different traditions of ecology, broadly defined, such as meteorology, fisheries, forestry, soils, hydrology, etc.

The trends sketched in Figure 8.2 have the following five implications:

1) From the early stages of economic development, where questions of resource potential and unpleasant aspects of the natural environment predominate, the data requirements have gradually shifted to what is needed to understand, order, and manage man's demands and impacts on specific resources and the environment.

2) Single factor, simple resource, and separate discipline conceptualizations are supplanted by more comprehensive initiatives that also address complex interactions, using functional analyses of dynamic man-resource systems.

3) Elementary scientific and technical methods appropriate at early stages are superseded by highly sophisticated and complex, precisely-ordered procedures, together with higher levels of concept coordination.

4) Simple records stored in survey notebooks are eventually succeeded by

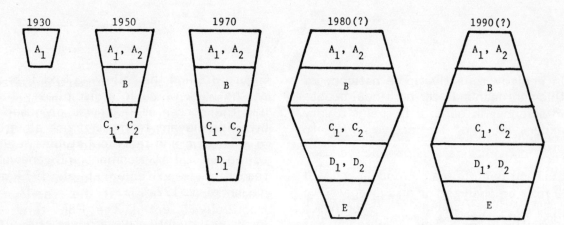

Figure 8.2 The size of the various strata is a very approximate estimate of the effort (corresponding roughly to the cost to Canada) expended in collecting, processing, and applying corresponding data classes.

A_1: An inventory survey, undertaken by workers in a particular narrow discipline. A few factors and quantities of immediate practical interest are measured on selected sites over a broad area. Development of specific resources may be initiated on the basis of such data.

A_2: Assessments of the potential of natural resources, whether renewable or nonrenewable, or of skilled manpower, etc. This step is used (1) in early stages of planning developments that require large social and/or capital investment and (2) in regulating resource use.

B: Assessments of the probable environmental impact of proposed large-scale developments. "Impact statements" are used in simplified socio-economic planning, e.g., in applying cost-benefit procedures.

C_1: Routine monitoring of indices, the mathematical definitions of which are derived from experience and scientific theories, and of factors that relate closely to the well-being of society from cultural, political, economic, or environmental viewpoints. Indices that measure fairly comprehensive factors over a broad area

may be useful for longer term projective/extrapolatory planning by government agencies.

C_2: Process monitoring and real-time control of some particular, important man-resource interactions such as the off-loading of noxious wastes or the harvesting of certain sensitive renewable resources. This approach tends to forestall excessively intense stresses from being applied within intensely interactive situations.

D_1: Step-wise, experimental management in which an explicitly formulated decision process uses information from past management experience and experiments, together with measurements of indices routinely monitored to guide future management.

D_2: Formal functional analyses of components of man-resource problems as dynamic systems. Usually a number of disciplines collaborate. This phase is still largely experimental.

E: Integrated planning and decision making using formal analyses of policy alternatives performed as much as possible in a transdisciplinary context.

data systems in a national network of electronic computers.

5) The more highly a region is developed in a conventional "economic" sense, the more intense the man-resource and the man-environment interactions are likely to be, and the greater the need for sophisticated transdisciplinary procedures.

By 1990 another layer, F below E, may have been added and developed to an operational state. It may involve some explicit quantifiable kind of public participation, e.g. public opinion polls, Delphi processes with random subsets of society, etc. (see Hetman 1973 and Benington and Skelton 1973). A wide variety of trials along these lines are already underway.

8.5 Interdisciplinarity Revisited

What really are some of the operational aspects of fostering interdisciplinary transfer sciences? This question may be addressed analytically from the viewpoint of Figure 5.1 from Jantsch (1972). Consistent with that viewpoint we may examine what is involved in "problem solving capability" (see Regier et al. 1974).

Problem-solving capability (PSC) may be perceived as being a function of: technical competence (TC) deriving from education, training or past experience; existence of appropriate theory (AT) or relevant models and methods derived in the most part by others in the discipline(s) of interest; and accessible data (AD) already in store or measurable in the field. Symbolically:

$$PSC = f(TC, AT, AD).$$

Much of science is compartmentalized into disciplines each with a number of "traditions" or "schools". Within a particular tradition TC, AT and AD tend to evolve as a co-ordinated and mutually congruent set with respect to scope, scale, special techniques, model characteristics, etc. If the tradition is practically oriented, the PSC tends to come as a prepackaged mechanism for practical application.

Identify a particular discipline by the subscript i and a tradition within a discipline as j. One approach to generalized problem-solving would be to identify and select the most effective and efficient among the available PSC_{ij}'s. This results in selection of the best discipline-tradition for a particular job: it ends as a unidisciplinary approach following competition among a number of disciplines and traditions.

Though there may have been little explicit consideration of alternative disciplines and traditions, the above approach has tended to dominate in fisheries studies, with "fisheries biologists" of the locally dominant tradition addressing many problems virtually in isolation.

Quite recently some aspects of fisheries concerns have been addressed by resource economists while fisheries biologists have continued to hold sway on other apsects. Occasionally sociologists and political scientists have assisted decision makers. Seldom has any effective cooperation developed, and the overall approach could be describe as being multidisciplinary (Figure 5.1). Symbolically:

$$PSC = PSC_{ij} + PSC_{kl} + PSC_{mn} + \cdots$$

where the effective PSC for synthesis may reside in undifferentiated form in the mind of some senior decision maker.

In water resource development studies, particularly in the early 1960's, benefit-cost analysis was the dominant methodology in a cross-disciplinary approach. Though little more than an accounting methodology, benefit-cost analysis was used in a rather coercive attempt to relate most if not all social and ecological considerations into monetary units to determine whether a proposal would be profitable if executed. Symbolically this differs little from the unidisciplinary case except that no explicit selection process was undertaken. The particular PSC was specified from the outset.

A major problem may be identified in terms of a broad concept, perhaps in rather philosophical terms (see Chapter 2). Conceptual screening of various discipline-tradition candidates may result in the identification of a set of these from different disciplines such that they are each congruent with the transcending concept used to specify the problem. If all members of the set are concordant (i.e. differences in semantics and conventions do not raise insuperable barriers), if the set does not exhibit major redundancies (i.e. overlap of linked dimensional components), and if the set is sufficient (i.e. possesses at least the minimal necessary dimensionality), then interdisciplinary cooperation might ensue.

Even though a problem may not be phrased initially in terms of a transcending concept, a number of workers in different disciplines who are collaboratively inclined may have found through experience that their approaches are conceptually congruent and compatible. Implicitly they may be addressing a common transcending concept but that concept itself may not be of any particular interest to them. They may recognize some degree of common-

ality in their concepts, methods and data (see Table 4.1). There may be a number of such mutually congruent sets, each containing one or more representative traditions from different disciplines. For example there may be one set that subscribes to a one-way causal paradigm, another set that subscribes to the random process paradigm, and a third set to the mutual causal paradigm (see Table 2.2). Thus, the task for the high-level problem solver is to select an optimum from among these interdisciplinary problem-solving capabilities.

The transdisciplinary approach as defined in Table 5.1 is more complex by an order of magnitude than the interdisciplinary approach. Values and norms that cannot be specified operationally for scientific purposes may well be implicated in problem situations where such an approach is initiated. Different discipline-traditions may contribute to the solution in a way analogous to the role of musical instruments in a symphony.

8.6 What's so Wonderful about Quantification?

B.M. Gross, though himself much involved in extending quantitative methodologies, recognized that "monetary information tends to drive out of circulation quantitative information of greater significance and quantitative information of any kind tends to retard the circulation of qualitative information" (Gross 1966).

Ecologically oriented scientists working with environmental and resource issues long ago recognized the validity of the first half of Gross's statement. Yet many if not most professional ecologists have made further quantification their primary objective. What is the tradeoff? What are we sacrificing in so doing?

G. Grant (1976), a professor of religion, has recently examined the implications of the statement, "The computer does not impose on us the ways it should be used". The statement was made by a man who worked at making and using computers.

Among other things Grant argues that the perception or paradigm of what it is to be "reasonable" both leads men to build computers and also to conceive the highest political goal as being a universal and homogeneous society. He continues:

The ways such machines can be used must be at one with certain conceptions of political purposes, because the same kind of 'reasoning' made the machines and formulated the purposes. To put the matter extremely simply: the modern 'physical' sciences and the modern 'political' sciences have developed in mutual interpenetration, and we can only begin to understand that interpenetration in terms of some common source from which both forms of science found their sustenance (p. 124).

When we, as western people, put to ourselves the question of what can be 'beyond industrial growth,' we are liable to be asking it as a problem within the package which is that destiny (i.e. within the conventional worldview, paradigm). However, even at the immediate level of the pragmatic, the questioning in 'beyond industrial growth' begins to reveal the (paradigm of) technology. We move into the tightening circle in which more technological science is called for to meet

the problems which technological science has produced. In that tightening circle, the overcoming of chance is less and less something outside us, but becomes more and more the overcoming of chance in our own species, in our very own selves. Every new appeal for a more exact cybernetics means, in fact, forceful new means of mastery over the ... lives of masses of people (p. 129).

Grant then calls on *thinkers,* as opposed to *practitioners,* to expose the current darkness as darkness and to refuse to participate in solving problems within the current paradigm.

Some of Grant's statements bring to mind Maruyama's indictment of the one-way causal paradigm (see Chapter 2). Also the writings of Albert Schweitzer, Jacques Ellul, Lewis Mumford, Ivan Illich, and many others come to mind.

For many thinkers the sequence depicted in Figure 8.2 might reflect the trend of defeat rather than of victory. My suggestion in Section 8.5 that the next major step in quantitative sophistication might involve something like "standardized methods of public participation" might be viewed with horror.

Perhaps our "information services" of the future might be better balanced and more useful if the content and trend depicted in Figure 8.2 were brought under more effective social control. The best things in life have no numbers associated with them.

9

the challenge for fisheries science in canada

9.1   Introduction

In Canada, as elsewhere one era in fisheries is ending and another is beginning. Effecting a transition has been and continues to be a challenge for the 1970s.

All major aspects of Canadian fisheries issues have been in deep turmoil during this decade. Because of intensified concerns about the natural environment, renewable resources and a variety of related issues and opportunities (see below), we have seen: upheavals in the fishing industries; major policy reorientations (FMS 1976); administrative restructuring; the maturation of policy and program planning at international, national and regional levels; and planning for science. The focus here will be on the last, but it cannot be treated in complete isolation of those other major changes that will determine to a large extent the kinds of science that will be needed and will be funded. Hence the chapter begins with a consideration of some broad perceptions of what is involved in the current historic transformation.

This chapter is based on work undertaken by a team under the auspices of the Fisheries Research Board of Canada

(1977) in its recent role as advisor to the Minister of Fisheries. The focus here is on the challenges and problems with the shelf-seas fisheries (Regier and McCracken 1975). Other fisheries-related studies of the team, that operated under the immediate direction of J.R. Weir and W.R. Martin, concerned specifically the near-shore marine fisheries (McCracken and Macdonald 1976), rehabilitation of anadromous and Great Lakes fisheries (Loftus 1976), the scattered fisheries of the Canadian interior (Regier 1976), environmental quality from a fisheye view (Harvey 1976), and aquaculture (Pritchard 1975). Though the *substance* of issues and challenges differed markedly with respect to these major sectors, the overall *methodological science framework* and *policy analyses* had much in common.

A consideration of the broader scientific framework is most relevant to the purposes of the present chapter. Using the shelf-seas issues as the practical setting for the discussion obviously introduces a bias, which can be corrected by referring to the other studies, all of which have been published in the Journal of the Fisheries Research Board of Canada.

9.2   Grotius' Assumption

Garrett Hardin's catch phrase "the tragedy of the commons" has come to be widely associated with the "common property--open access--willing consent approach" to the natural environment and renewable resources (see also Ostrom and Ostrom 1971).

With important local exceptions, total fishery catches continued to expand through the 1960s as though consistent with Grotius' 17th century assumption that the seas contain an unlimited mass of harvestable living resources (Finkle 1974). This growth was achieved to an important degree by expanding the harvest process regionally, finally including the farthest reaches of the oceans. As the intensity of the harvest processes grew locally, certain stocks of whales, clupeids, gadids, salmonids, crustaceans, and molluscs were suppressed to be replaced ecologically by less valued forms. The abundance of the lower valued replacement species was clearly not vastly greater than that of the preferred stocks at the peak abundance. If this had been the case, high volume, industrial fisheries might profitably have replaced lower volume food fisheries. Total harvests have fallen in some areas where fishing was very intensive and/or environmental factors have changed.

It is now universally recognized that Grotius' assumption has outlived its usefulness. The new law of the sea is clear testimony to that. The exuberance and excesses of a fishery process that perceived large resources still untapped are being replaced by coordinated deliberate actions. Resources must be husbanded (Davis 1973). A system of negotiated allocations is gradually displacing opportunistic overharvesting in some regions. Where, in the past, market demands for fish protein have often been weak, we now expect demand to exceed supply for some decades to come. Instabilities in the form of local outbursts and collapses will likely intensify. The world demand for food is expected to double by about the year 2000 (FAO 1974). It is unlikely that the world yield of preferred marine fish can be doubled--perhaps it can be increased only by 50% over present catches (Needler 1973).

The concerns of conservationists are no longer dismissed. In the past some of the conservationists' warnings may have been premature. Or they may mistakenly have been judged to be premature by fishing interests that could shift from an overfished stock to an underfished resource some distance removed, but still maintaining pressure on the overfished resource (Larkin and Wilimovsky 1973; Lawrie and Rahrer 1972). The possibilities of low cost frontier or nomad exploitations are rapidly approaching zero. When all resources will have been "fished up" further increases in the value of catches per se--aside from inflation and shifts in consumer preferences--will stem from improvements in various aspects of management and harvesting broadly defined. Improvements will involve ecological, technological, economic, and social initiatives.

Laissez-faire in fisheries is now passé. What was held to be irrelevant or politically inaccessible is now being addressed directly, e.g. explicit allocation in a common property resource. Some central presuppositions are now discredited, such as Grotius' concept of the freedom to fish anywhere. In the short run this complicates everything because much of our thinking has to be reorganized. But we may expect that needless complexity will dissipate as a new paradigm or policy becomes well established (Chant and Regier 1972).

9.3 Fundamental Perceptions and Decisions

Previously, we asserted that the challenge of the 1970s is to terminate one era and to initiate another. This applies to environmental and resource matters in general and thus extends beyond fisheries. We look particularly to those decisions implicating scientific research related to fisheries and the primary fishing industry or users.

a) That *renewable resources are limited* and that the overall supply cannot readily be expanded by the ingenious and inexpensive technological fix is now broadly accepted (FAO 1974). The technological optimists who trumpeted overblown hopes for 1000 million tons of marine protein or an imminent "green revolution" are now being ignored. Largely ignored also are the few sectarians who still threaten with postmortem pain those scientific workers who advise some birth control, and who urge instead that scientists dumbly and faithfully help to feed "all the dear children that it pleases God to create".

Canada's initiatives at the 1972 U.N. Conference on the Human Environment, at the 1974 U.N. Conference on World Population, at the 1974 FAO Conference on Food, at the Third Law of the Sea Conference, and elsewhere, clearly imply strong international commitments and responsibilities on renewable resources. In Canada reserve productive capacity for food is rapidly being taken up in response to increased national and international demand. Large amounts of productive capacity are still being sacrificed to urbanization, waste loadings, pollution, and careless production practices that are widely recognized as undesirable but are continuing almost unabated.

In Canada there has been little scientific interest in larger questions of international allocation and equity. The old-fashioned solutions of various denominational and humanitarian charities threaten to be self-defeating because ecological and other dimensions of the problems are generally ignored by these groups. (Some new, more balanced initiatives should not be dismissed here.) Meanwhile we feed wastefully on abundant resources. Our personal and national security is not well served by ignoring the larger issues of worldwide resource scarcities.

Clearly extension of management sovereignty over all our shelf seas will bring us into close involvement with other nations addressing the Gordian knot of the growing shortages of animal protein worldwide and the growing waste of protein in Canada and in other developed countries (Borgstrom 1974).

b) Grotius' enunciation of the *freedom of the seas*, in particular of free access to the fisheries of the seas, is no longer valid. Everybody's property is nobody's; when the technological and enterpreneurial capability exists to harvest all stocks, powerful regulatory or allocative mechanisms must be established to prevent overharvesting. The Canadian government has in effect committed itself to solve the "common property--open access problem" (Davis 1973). Progress is slow.

With effective fisheries regulation the demands for scientifically respectable information will become intense. Within a common property--open access regime the practical role of science is little more than scouting and assessing resources in the vanguard of the expanding process of further "development". Such activities as specifying minimum mesh sizes, minimum legal sizes, open seasons, closed areas, etc. are singly or jointly ineffective if the regime is basically one of common property--open access. If in addition "willing consent" implies unanimous acceptance of recommendations that will have some unpleasant effects in the immediate future, then demands for information become excessive. Much information is needed in the conventional process but there is little scope for effective action. This has long been sensed by scientists, with a result that "practical fisheries research" has seldom held for long the commitment of creative scientists. Fortunately for both the industry and scientists, those days will soon be past.

c) Canada's motives for seeking to *extend management sovereignty* over our anadromous salmon wherever they may wander, and over the resources of the shelf seas and beyond to 200-miles where shelf seas are narrow, include the argument that our inland and onshore activities markedly influence offshore fisheries (Haig-Brown 1974). Because we must forego benefits on land in order to protect offshore fisheries, we deserve special rights that can be assured by a grant of management sovereignty. What these inland and onshore sacrifices are, and how they relate to offshore fisheries will need to be spelled out eventually if a coastal nation is to retain credibility.

More than that, it is generally held that a nation's scientific corps, as well as the managerial and diplomatic, will need to be first-rate if it is to assist effectively in resolving complex issues. The lead role in resolving difficult issues, now usually shared by a number of nations, will probably come to rest more squarely on the coastal state.

Whether from the viewpoint of nationalist advantage of transnationalist stewardship or a synthesis of the two, it is clear that a coastal nation's scientific corps must be developed so that it will not suffer in international comparison.

d) *Allocation*, using a complex system of quotas based on biologically estimated total allowable catches (TACs) was initiated and applied by ICNAF. Interest in explicit allocation, ultimately with respect to individual fishing enterprises, is growing within Canada. Such a policy might greatly simplify programs of rationalization, i.e. the achievement of economic efficiency (Mackenzie 1973). Municipal, provincial, and regional interests will participate in such a process.

e) The owners of a resource generally expect some benefit, *resource rent,* from it. We are coming to expect that the direct user pay resource rents and that the abuser, direct or indirect, pay penalties for harm done. In Canada the direct users of the fisheries resources have often been subsidized. This may be viewed as seed money, or wealth redistribution to compensate for regional inequities due to other national policies, or a consequence of pork barrel politics, etc. Whatever the inequities of the past, there is a growing interest in generating resource rent, at least from the more valuable and lucrative fisheries resources. Internationally the question arises as to who will pay for research and management costs where multinational fleets use resources for which a coastal state will exercise management sovereignty.

License limitation or entry control together with explicit resource allocation may well be an effective way to solve the common property--open access problem in order to recover resource rent through taxes, royalties, etc.

With a new regime, fishery enterpreneurs, whether private or state, and managers may recognize relevant scientific information as having primary usefulness. No longer will scientists' involvement be restricted to peripheral questions of mesh sizes, seasons, etc. They will now assist in finding appropriate fishing procedures and regulations to rehabilitate fisheries, maximize resource rents or to achieve some other target.

f)   The environmental or *ecological revolution* is the complementary concern to that of renewable resources from a socioeconomic viewpoint. The natural environment is the habitat in which those resources generate themselves. It may be considered a "resource" in that it also accepts and inactivates or regenerates the many wastes we must dispose.

Jointly the concern for habitat health, living resources' productivity, and sympathy for certain taxa such as baby seals and whales, are merging with other forces in society and are being incorporated within new institutions. These forces have been led in part by scientists, and the expectations that such interest groups hold concerning balanced science are more than moderately great. They will expect that major problems be examined in larger political and technical contexts than has been done commonly heretofore.

Part of the resource rent that a nation seeks to recover from a natural ecological system may well be harvestable only as aesthetic, emotional, or ethical satisfaction. With growing populations and intensifying demands on resources, such matters can only become more complex, if allowed to develop in a haphazard manner.

g)   Nationally we are moving to *counter extreme manifestations of urbanization,* in which relatively few urban areas are growing at rates beyond all reason. Hence the interest in viable small- to medium-sized communities, alternative life styles, labor intensive enterprises, the service industry, and small establishments (Macpherson 1972). Politics of economic rationalization will need to meet the test that they do not contribute to further undesirable urbanization.

Clearly there are needs for research by political scientists on institutional infrastructures and policies. Sociologists, geographers, and the disciplines more usually associated with fisheries--economists, engineers, fisheries biologists--will need to work with political scientists in transdisciplinary contexts. Because the scientific competence is now accessible regionally it would be inexcusable to blunder through this process of redirection purely by trial and error or through piecemeal administrative fiat by politicians often striving competitively against each other.

h)   As our social fabric has become more convoluted and interwoven, demands for *explicit planning* have been heard and to some extent heeded. These demands arose in industry, in government, and in various institutions: they are not only seeking to enhance personal security. One of the common tests of planning efficacy is whether or not major alternatives to the status quo are identified and analyzed in some detail.

When common property-open access-willing consent regimes were dominant, it was held that freedoms were not to be constrained except on particular scientific evidence so strong that it commanded consensus even of parties strongly biased against such advice. With a displacement of earlier policies by those of licensing, limited access, quota regulation, and experimental management, scientific information is needed for direct action as opposed to the more constrained role of indirect reaction earlier.

It has often been stated that effective regulatory protocols can be developed on a semiexperimental basis that does not require full scientific understanding. This is quite true. On the other hand, so-called experimental approaches, that are little more than undisciplined trial and error, would again lead to great waste and inefficiency.

The basic shift in development and management approaches does not call for less science but rather calls for the type of science appropriate to the new way of doing things. The responsibilities for broadly based science increase and intensify, and the necessary information becomes a primary basis of decision, rather than a peripheral and ineffectual constraint, as before.

i) Within Canada there appears to be a growing commitment to *cost-effective, interdisciplinary "big science"*.

In their own rights, science and technology have become foci for intense policy debate in Canada and elsewhere. Various studies have been performed under the auspices of the Science Council of Canada, including several that relate to fisheries problems. Modification of the role of the Fisheries Research Board is just one event in many related to policy changes.

With continuing commitment to plan rationally for science and technology, the case for fisheries science and research will need to be made explicitly in a forum where other interests will also be heard. One clear implication is that scientific prima donnas will often find a frosty welcome in research budgeting(Hayes 1973), and only carefully justified research proposals will be funded. The case for particular studies will need to be made clearly and intelligibly to senior officals and others not fully conversant with the theoretical scope and details. This will demand skills that have not heretofore been seen as necessary in Canada, especially not within the government scientific establishment. American academic scientists have had to be far more specific in these respects than was the case in Canada until very recently.

j) The research planning, programming, and funding procedures with respect to Canada's fisheries resources are at present relatively unorganized. A national research institution, the Fisheries Research Board with a network of laboratories and granting functions, with all its shortcomings was recognized for its excellence around the world; it has been taken apart and the pieces have not yet been put together in an alternative viable structure. A number of senior scientists recognized the world over have recently retired from service, and the succession has remained unplanned. Meanwhile a university grant program has been transformed into a process in which university scientists are expected to compete to serve as junior assistants to younger scientists in the federal service. Add to this a government make-or-buy policy which is intended to stimulate private enterprise to enter the fields of science and technology on contract in the hope that science will be more effectively transferred into technology. The first decade of the science that is produced under contract to private enterprise can hardly be expected to be first-rate, though it may be useful science (Meyboom 1974). Thus at a time when Canada badly needs effective science on fisheries resources, the country finds itself without an appropriate institutional structure to stimulate, guide, and achieve scientific success.

k) Quite clearly the methods of science, shared by all peoples of the world, are being mobilized in part to forestall confrontation by force. This can work only so long as scientists are technically competent and objective. Special consideration needs to be given to the latter proviso.

With respect to renewable resources and the environment, institutions of various levels are rapidly being created, largely in an ad hoc manner, and are being tested. No longer can a few senior bureaucrats, politicians, industrialists, and fishermen make major decisions quietly on the simple basis of their shared, informed judgments. It is becoming much more complex than that. Predictably each major interest or group that is involved in the planning and decision making will wish to have scientific assistance relevant to its purposes.

Even under conditions approaching total freedom, scientific objectivity is an ideal that is difficult to realize (Mumford 1970). The ideal remains remote where scientists must offer scientific advice to their personal superiors who also bear direct short-term responsibility for practical and political decisions. Under such circumstances personal job security often takes precedence over a commitment to objective science.

Concluding this section on policy perspectives, what has been elaborated above may be condensed as follows. Various fundamental trends and decisions have led Canadians to view their shelf fisheries as of greater significance than heretofore. This has come at the same time when Canadian governments are experimenting seriously with policy and program planning in a rationalist mode. One of the policy areas that has received intensive consideration is the new planning approach in science and technology. The main effect to date is that the institutions in place in the 1960's with responsibilities for scientific research in fisheries are no longer effective, and new ones have not yet been well established. To continue to drift is to risk the total collapse of fisheries science in Canada.

9.4 Meeting the Challenge: Organization, Programs, Facilities

During recent years many analyses of research needs in fisheries have been published; in Canada these include works by Pimlott et al. (1971), Larkin et al. (1970), Stewart and Dickie (1971), and others. The present proposals are generally consistent with those of others but have been further developed.

The following proposed priorities for fisheries science are in general not inconsistent with what most other groups in the world are now thinking. For example, these proposals have been compared with many papers of the 1973 FAO Technical Conference on Fishery Management and Development (Ricker and Weeks, 1973). But there is strong inertia with respect to change.

The tradition that developed with respect to North Sea circumstances in the 1950's became dominant in FAO with a result that technical manuals that would have been particularly relevant to the North Sea at the time were produced and promoted worldwide. The tacit assumption appears to have been that what was good for the North Sea must be good for the world. To a limited extent the assumption was valid. Originators of this program in FAO have long since seen its limitations and have initiated new directions (Kesteven 1972). Also, FAO officers such as J. Marr (1973), M. Ruivo, S.J. Holt, and J.A. Gulland have stimulated a search for alternatives. Yet the narrower preoccupation is still too much in evidence in committee reports (see Lucas et al. 1974). Here are some proposed priorities to improve fisheries science to meet the current challenges.

1. Priority on mapping: Consolidate within regional libraries and periodically update important distributional information on resource habitats, resource stocks at various life history stages, harvests, harvesting capability, and other socioeconomic quantities.

Oeanographers and geographers have developed comprehensive regional map libraries. Some universities and government agencies now have compendia of many thousands of maps. Relevant expertise should be coopted to find and bring together existing fisheries maps and to integrate them into regional map libraries.

A program of updating old maps, where they exist, and developing a balanced series of new maps should be undertaken expeditiously.

Photography, from survey and surveillance aircraft and high level satellites, is making available a vast number of "maps" essentially on a real-time basis. Technical specialists who can efficiently access such information for the advantage of fisheries management and science should be attached to the regional map libraries. Some research on mapping methods will be necessary on a continuing basis.

2. Priority on monitoring: Within regional fisheries offices, consolidate and then maintain in current readiness, series of monitored data on ecological, economic, and social factors of particular practical relevance for the 1980's.

For the Atlantic fisheries, organization of monitored data series on oceanic variables, sonic and trawl surveys, catches, harvesting effort, etc. has been stimulated by the need to estimate a large number of "total allowable catches" or TACs. On the Pacific coast progress along these lines has reached different stages within the various international agencies. On both coasts, the current status can fairly be labeled heterogeneous and a strong commitment toward rationalization is apparent. Fisheries monitoring units' terms of reference should explicity recognize the probability that 1975 arrangements in ICNAF requiring a large number of TACs to be estimated annually, will probably be superseded within a decade by alternative approaches. These new approaches will probably also be introduced off the west coast; if so, the 1975 mix of ad hoc international agencies may be replaced by one or two more comprehensive and effective arrangements. This implies that the fisheries monitoring units have an advisory panel of the more innovative managers, resource scientists (ecologists and economists), and political scientists who will assist in identifying important variables to be monitored.

The mapping and monitoring units might be brought together regionally in scientific services centers. In any case, organized separately as units or as two components of a center, their terms of reference should be based on a review of recent experiences of meteorological and hydrographic services.

3. Priority on management and harvest protocols: Consolidate regionally the scientific and support staff necessary for calculating TACs and eventually for specifying the kinds of management and harvest protocols that probably will replace the present system based on TACs.

This activity, to the extent that it is based on scientifically valid insights and information, is clearly a management function. Technology transfer, i.e. the application of new science, is involved where ongoing science is actively contributing to the improvement of the harvest regime. Where little is known scientifically, or the conventional science is not highly respected, scientists should participate directly by helping to organize harvesting protocols as experiments, assuming that appropriate analysis and evaluation can be assured.

Management and protocol centers should be established on each coast to assume the responsibility for marshalling information necessary for all the important formal negotiations concerning fisheries harvests. Assuming that the three "shells" of policy and decision making will evolve with respect to national marine fisheries interests (i.e. open international ocean, shelf seas under national management, and inshore national seas), the management and protocol centers will need to be responsive to major needs in all three "shells." Clearly the mapping units and the monitoring units will provide information to the management and protocol centers but the former should not be formally and fully integrated into the latter. Close liaison with fisheries research centers (see next priority) must be cultivated. Provincial (or state) fisheries agencies should be invited to second experts to collaborate with federal counterparts in management and protocol centers.

The overall system must be kept flexible to accommodate future changes. If the units and the centers were formally integrated, it is unlikely growth of a strong resistance to further change could be prevented.

4. Priority on synthesis and modeling: From the perspective of transdisciplinary resource science a number of fully viable fisheries research centers should be established. Through synthesis and modeling the centers should seek to understand, simulate, and explain major natural and man-made events and processes. Models that encompass social, economic, ecological, and technological components should seek to deal inter alia with market mechanisms, social infrastructures, macroecological processes, and technological innovations. Such science would facilitate prediction of consequences of various management options concerning fish and other aspects of interest in aquatic systems. Also, it should find effective ways of measuring performance of various government programs.

Fisheries research centers focusing on marine systems should be closely linked with similar centers studying freshwater systems. A concerted effort should be directed toward effective interaction between freshwater and marine workers. Some frameworks now exist within which insights and information from both can be brought together to great mutual advantage.

Broadly speaking the scientific approach should be consistent with the way in which Canada's Man and the Biosphere Program is developing. The identification or development of promising hypotheses, from a transdisciplinary perspective, the subsequent testing in simulation and experimentation, and the application of firm inferences, are the core of that process.

Much of the nation's fisheries research activities should be explicitly related to the transdisciplinary programs undertaken by the fisheries research centers. Aspects that might not readily be integrated include fully innovative and curiosity-directed research as well as attempts to reconsider some earlier approaches that have been neglected.

5. Priority on facilitation of experimental management: A capability for practical, planned experimentation should be incorporated within the system of programs and agency infrastructures that is being developed as an alternative to the obsolete "common property-open access-willing consent regime." The assistance of political scientists may well be essential in order to achieve such a capability.

Transdisciplinary experiments should test hypotheses derived from currently best information and insights available from the mapping, monitoring and modeling components of science. Such an experiment should be a collaborative effort between a center for harvest protocols and a center for fisheries research plus assistance from mapping and monitoring units. The center for harvest protocols should take the lead.

It is important that the "experimental management" not become a euphemism for ad hoc crisis responses or a trial-and-error "blundering through" process. The latter tactics may on occasion be fully justifiable, yet they should not be graced with the adjective "experimental."

6. Priority on division of labor: Develop appropriate screening procedures, incentives as well as successional and transfer programs for three major classes within the scientific manpower mix: for those whose work is program-dictated, or is mission-oriented, or is curiosity-directed.

None of these classes should be considered of inherently higher status to the overall interests of a nation's fisheries. Yet because of heavier demands in training, inertia against professional and educational innovation, etc., it may be that manpower shortages in one or another class will occur; additional recruitment incentives may then be mistakenly interpreted as implying higher status. Possible difficulties of this sort may be minimized by developing programs of upgrading and retraining for the benefit of ambitious personnel.

Many technically expert workers do not aspire to any major creativity and some would break down under any coercive stimulus in that direction. Others are highly creative conceptually with hardly a practical bone in their bodies. Some like to fit together pieces of a scientific puzzle to detect broad patterns even where less than half of the pieces yet exist. Within the mix of essential work, appropriate tasks for each group can readily be identified.

The brunt of implementing the priority recommendations here listed--assuming they were to be found generally useful--should fall primarily on the class of mission-oriented scientists who have achieved some transdisciplinary freedoms. Appropriate roles for the mission-oriented would include those of chief of regional mapping and monitoring units, as well as some of the senior positions, including planning and programming in the regional protocol and research centers.

Scientific and technical experts committed to program-dictated activities would perform most unidisciplinary components within the programs of the mapping and monitoring units and of the protocol and research centers, and would also assist in the work of self-directed scientists.

Retrospective or curiosity-directed scientists would usually be senior or exceptional scientists who have through good works or obvious creative talent won the freedom and funds to reinterpret and reassess past work and to explore uncharted waters. These should be associated with regional research centers, cross-appointed to universities, and generally allowed about as much freedom as possible. One of the important constraints on their freedom should be a rigorous system of peer reviews of all study proposals. The day of the blank check is gone, at least for a few decades.

Whether a scientific worker's activities are program-dictated or mission-oriented, some significant stimulus for personal initative and scientific creativity should be built into the reward system. Curiosity-directed workers' reward systems will of course rest very squarely on their creativity. Opportunities for transfer should be based on achievement of appropriate upgrading and retraining.

7. Priority on planning, performing, and reporting research: Making explicit a thread woven through the priorities already specified, research activities should be mobilized into programs with clear objectives adequately reviewed by peers, conducted according to planned schedules, and reported expeditiously.

Accountability criteria applicable to individuals should include one that specifies that new or continued funding of a scientist responsible for a study should depend on satisfactory reporting of previously funded work within a reasonable time; exceptions should be granted only by the senior officer of a center or a region and be duly recorded. Too many fisheries workers of recent decades have been permitted to neglect their duty to report adequately on their work.

On a larger scale, accountability of the total regional research center should remain a concern, and methods of assessing overall effectiveness should be developed further and applied routinely.

Research planning and decision making should be undertaken with respect to transdisciplinary perspectives. As soon as this can be achieved, workers originally trained in different disciplines will become more collaborative and collegial, rather than competitive and adversary as is so commonly the case in recent interdisciplinary attempts. Individual scientists who understand the paradigm of only one narrow discipline, particularly if they are also unilingual or unicultural, could not be expected to contribute centrally to the overall planning and guidance of the research system.

8. Priority on scientific credibility: Clarify the sociopolitical roles of various governmental and other research agencies and institutions, scientific societies and professional associations and then if necessary develop countervailing mechanisms and processes to guard against the subversion of fisheries science for national or other political or personal purposes.

National interests, the short-term self-interest of scientists with strong personalities, or an uncomprehending conservatism, all can lead to the subversion of science. Recent controversies on questions such as the relationship of environmental radioactivity to cancer, smoking to respiratory diseases, phosphate detergents to eutrophication, fishing intensity to species collapse, etc. show clearly that scientists can differ. Undoubtedly scientific objectivity can be bought off or rationalized away in the service of personal, or national goals. There has been much muttering about this problem; it needs now to be faced creatively.

An international scientific society, in which fisheries science is argued on its merits and is not likely to be readily influenced by political interests, should be considered for both the North Atlantic and the North Pacific. Existing institutions, such as the International Council for the Exploration of the Sea, the Pacific Science Congress, and periodic symposia sponsored jointly by a variety of international bodies do not adequately take the place of mature scientific societies.

10
mobilizing the scientific/educational/innovative system

10.1 Why Mobilization?

In recent years issues related to renewable resources and the natural environment have come to be recognized as of primary importance. This has involved a major reorientation of social perceptions. Money and manpower resources have been directed from other earlier preoccupations into environmental and resource "management", very broadly defined. Many of these resources have flowed through the accounts of enterprising individuals, corporations, institutions, and agencies that understood little about how ecological systems work but did understand how to mobilize into reassuring activity and had few compunctions about spending large amounts of money.

The environmental impact assessment process and programs to develop environmental monitoring systems are examples of the above. The great majority of ecologically sophisticated people were either too busy trying to become more sophisticated or were preaching environmental doom. When money was made available, well organized groups who had earlier benefited from the arms race and the space race had no

difficulty capturing large chunks of that money. Perhaps some of them have learned about ecology since then, but as an educational process it may have been very costly.

Is it safe to predict that such well-organized (but only semi-informed) groups will capture many of the scientific contracts and funds that will be made available to address the new challenges following the current Law of the Sea negotiations? Will the experts who feel that they know something about fisheries resources, aquatic environments and the harvesting and stress processes be so closely programmed into ongoing small-scale activities as to ignore most of the major scientific opportunities? If the current experts wish to get into the action more or less on their own terms, they will probably have to take some major organizational initiatives.

To mobilize implies seeking the power and means to achieve certain ends. The greater purpose to be served by the mobilization of fully fledged experts on resources and the environment is to solve major problems using models and methods that are relatively inexpensive. It may be more costly overall to assign a job to a poorly organized group of real experts than to a well organized group of semi-experts. But doesn't mobilization involve a sacrifice of the essence of science--creative discovery?

10.2 A Dilemma

Consider the following excerpt from Medawar (1976) on the scientific method:

However, there is an ignorance--amounting sometimes almost to contempt--of scientific philosophy not only among scientists but also among people, professedly critical thinkers, who ought to know, and often profess to know, better. This has led to the widespread misconception that the scientist works according to the rules of some cut and dried intellectual formulary known as "the scientific method." It has therefore come to be widely believed that given money and resources a scientist can bend the scientific method to the solution of almost any problem that confronts him. If he does not, it can only be because he is lazy or incompetent. In real life it is not like that at all. It cannot be too widely understood that there is no such thing as a "calculus of scientific discovery." The generative act in scientific discovery is a creative act of mind--a process as mysterious and unpredictable in a scientific context as it is in any other exercise of creativity.

We cannot devise hypotheses to order. Shelley would have understood this perfectly, for in his <u>Defence of Poetry</u> he wrote:

A man cannot say "I <u>will</u> compose poetry;" The greatest poet even cannot say it .

Nor can even the greatest scientist undertake to have illuminating ideas upon any problem he is confronted with, though he will know probably from experience how to put himself in the right frame of mind for getting ideas, and what reading and discussions will help him have them.

Also consider some metaphors by Waddington (1975, p. 246):

Man in the world is like a caterpillar weaving its cocoon. The cocoon is made of threads extruded by the caterpillar itself, and is woven to a shape in which the caterpillar fits comfortably. But it also has to be fitted to the thorny twigs--the external world--which supports it. A puppy going to sleep on a stony beach--a "joggle-fit", the puppy wriggles some stones out of the way, and curves himself in between those too heavy to shift--that is the operational method of science (and of the evolution of biological systems).

The pervasiveness of a concept of evolution, related to creativity broadly defined, is worth noting. Waddington (1975, p. 282) refers to the general philosophical implications of the theory of evolution as involving the substitution of dynamic for static categories of thought as well as an interest in origin and becoming rather than in mere being. This trend is coming to be felt as well on environmental and resource issues, perhaps somewhat belatedly.

Jantsch (1972, p. 218) distinguishes between discovery and creation:

Thus we find ourselves in a dilemma, from what angle of view . . . to try to elucidate . . .   science--God's or man's? What is man's principal task in dealing with science, perception or creation? Do we want Mozart or Beethoven--the image of man in God, or that of God in man--

Palestrina or Wagner--meaningful structure or structured, passionately individual meaning? Is choice really a necessity here? Is there a choice at all?

The answer is that there is no resolution to this dilemma, that it constitutes perhaps one of the basic paradoxes with which we must learn to live, and which give enhanced meaning to human life. The human condition *in the scientific technical era may, again, find its ultimate expression in the spirit of the ancient Greek tragedy, in which man attains his full creative freedom in making the "structures" (not simply the laws) imposed by the Gods his own for taking purposive action.*

Following the flourish on an ancient theme, Jantsch then committed himself as follows:

Organization for a purpose implies the introduction of normative and pragmatic principles which are beyond the traditional notion of empirical/conceptual science, which part of the science/innovation system is accepted under the name of science, and which is not, is totally unimportant. What matters is that science is recognized as part of the human and social organization.

The Science Council of Canada has recommended to the government of Canada that the highest priority for Canadian science and technology be assigned the task of helping to reorient Canada into a "conserver society". No doubt about the normative and pragmatic implications of that suggestion!

10.3 Science and Accountability

The current ferment on matters of science policy will not be easily resolved.

Science as discovery or as creation will continue to be scrutinized carefully to determine its relationship to our civilization's deeply held values and norms--recall the excerpt from Grant (1976) in Chapter 8. Science and technology separately and jointly may be held accountable to a greater extent than heretofore.

Yet it seems fair to say that two other approaches to "truth" other than the scientific should also be held more strictly accountable than is so now. "Truth through religion," with its massive denominational organizations and capital holdings, has anything but an unblemished record. "Truth through raw power" is an approach that sustains vast war budgets and precipitates much carnage. By comparison to these other approaches to consensus-building and truth, each of which commands vast social resources, science as science may still possess a very good record. If this is the case, then it would be unfortunate if science were to be held more strictly accountable while mighty militarists and God-fearing churchmen were permitted to continue unchecked in relatively irresponsible ways.

Churchmen often decry the close liaison that has developed between the military and certain fields of technology. Yet many Christian denominations themselves relied heavily on technology (e.g. medicine, agriculture) in their attempts to proselytize--even on the military itself on many occasions. Few missionary endeavors have been unmixed blessings and some have even involved genocide, directly or indirectly.

Therefore let us have fair social accountability, not only for science but for other organized approaches to extend the substance and process of major paradigms.

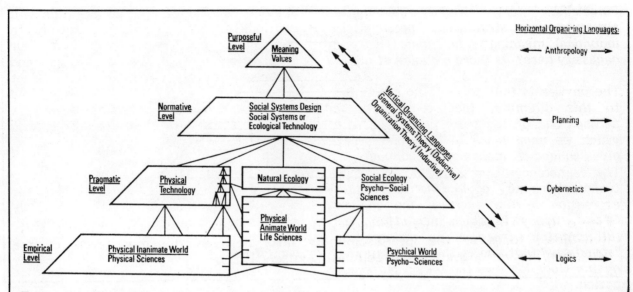

Figure 10.1   The science/education/innovation system, viewed as a multiechelon hierarchical system. Branching lines between levels and sublevels indicate possible forms of inter-disciplinary coordination (Jantsch, 1972).

Many scientists are actively reconsidering the presuppositions that have dominated theory and practice in recent years (see Chapter 2), as have many churchmen (Baum 1975). Militarists appear to be less questioning.

10.4 Science/Education/Innovation Systems

Jantsch (1972) has attempted to characterize a "system of science, education and innovation" (SEI System) that seems to be emerging in the West under broad social pressures for more "creative action" on the part of social institutions and agencies. (Again recall the works of Chevalier and Burns, 1975; Fox, 1975; and numerous others.) Figure 10.1 is a sketch of the conventional paradigm of the SEI System as it has evolved during the past century. If widely judged to be inadequate, it may be complemented in such a way as to make the new combined system more nearly adequate. If obsolete, it will presumably be replaced by a more useful alternative. Almost every major scientific, educational, and governmental agency has come against one or both hypotheses-- inadequacy and/or obsolescence--during the past decade. The inadequacy hypothesis has usually been taken up, perhaps because it is easier to add something to what already exists than to achieve a radical transformation. Here and there new educational institutions have been modelled on a concept alternative to that of Figure 10.1 and some experience papers are now becoming available (Jantsch 1972; Francis 1976).

Parenthetically, the term "radical" need not imply physical violence, though some violence directed toward abstract constructs is inevitable. The radical process here implied is one of creative evolution, not bloody revolution. The motivation for the latter type of trans-

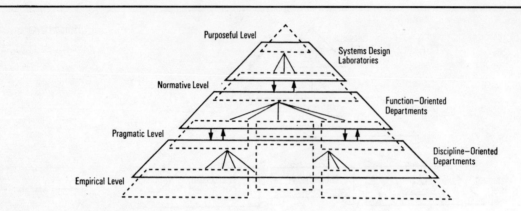

Figure 10.2   Transdisciplinary University Structure. The three types of structural units (full lines) focus on the interdisciplinary links between the four levels of the education/innovation system (dotted lines) (Jantsch, 1972).

formers may often be pathological rather than creative. It is unfortunate that a good term such as "radical" has been so commonly misapplied to pathological individuals.

In Figure 5.1, the bottom sketch refers to "transdisciplinarity". Its configuration resembles that of Figure 10.2 in an approximate way. But the hierarchic structure has in large measure been abstracted or disengaged from it. The strong arrows imply process orientation--a kind of compounded interdisciplinarity. This may be altogether too fluid for our present tastes, trained as we are to focus on structure and authority rather than on process and creativity.

Particularly intriguing is the central location in the pyramid allotted to "natural ecology".   Though not himself an ecologist, Jantsch apparently recognized that the perspectives of ecology transcended those of the natural sciences, that ecology employed a somewhat different "organizing language", and that ecological systems required pragmatic modes of intervention in normal practice.

Clearly Jantsch was more influenced by the holistic concepts of ecology than by the dominant reductionist methodologies-- a focus on the latter would have implied that ecology should be assigned to the empirical level in Figure 10.1.

Figure 10.2 depicts a kind of half-way step from the conventional wisdom (Figure 10.1) and a radical alternative (Jantsch's transdisciplinarity in Figure 5.1).   It derives from an attempt by Jantsch in the late 1960's to sketch an alternative strategy for the Massachusetts Institute of Technology (see Jantsch 1972, p. 235 ff.) in which:

The basic structure . . . may be conceived as being built essentially on the feedback interaction between three types of units, all three of which incorporate their appropriate version of the unified education/research/service function.

These suggestions may have been prophetic (i.e. seeing into the contemporaneous, existential present) in that a broad trend along these lines may be noted, even where few people have become aware of Jantsch or the numerous other planners working in a paradigm generally congruent with his. Thus many of the "interdisciplinary" centers, institutes, and programs of large universities and research institutions now may be assembled into the framework of Figure 10.2 with some gain in understanding. Parenthetically, the descriptive terms associated with these bands on the right slope of Jantsch's pyramid are not very informative, to my way of thinking.

The new instruments created by governments to deal with man-nature problems have often been more closely associated with these new bridging ventures in universities, and similar institutions in the private sector, than with the conventional disciplines as organized in Figure 10.1.

Note again the niche of natural ecology. It is shown as relating largely to a different bridging category than the rest of the "natural sciences". Thus ecology has been associated primarily with the pragmatic-normative bridge rather than with the empirical-pragmatic. This makes a lot of sense, since it captures in essence much of what has been addressed in previous chapters.

10.5 Prospects for Resources and Environmental Science

During the past decade western nations have developed or strengthened a series of instruments to correct the worsening man-nature interaction (Chapter 1). It need not surprise us that virtually all of these admittedly ad hoc initiatives have so far fallen far short of full effectiveness. Part of the reason may be related to the fact that the ecological input has seldom escaped a slide into the reductionist method. Thus Leopold matrices,

emission standards, agglomerated air or water quality indices, simple concepts of carrying capacity, maximum sustainable yield, etc., singly and jointly fail to address and safeguard the essence of ecological systems as perceived by holistic ecologists or holists of any disciplinary persuasion (e.g. Galbraith 1964).

What would be appropriate environmental methodology to marry with holistic concepts has hardly ever been addressed explicitly by ecologists. In order to be applied in practice, a holistic science of ecology--as may also be the case with holistic social sciences--may imply important changes in decision processes (see Bella and Overton 1972, and Emery and Trist 1973). The slowly emerging planning process may pave the way for more holistic approaches, particularly if it involves some public participation. Planning itself may gradually be turning from reductionist methods such as benefit-cost analyses or program planning and budgeting (PPB) into more explicitly dialectical methods (Benington and Skelton 1973). Or it may emphasize "creative learning" (Fox 1975).

Taylor (1976) may have had something like the above in mind when he stated:

. . . planning in a society like ours . . . involves more than just a coherent synthesis of many orders of consideration. It requires also the creation of a political consensus, or perhaps better, a resolution of widely felt ambivalences, broad enough and firm enough to sustain a plan over time. Planning is a very <u>political</u> activity: and yet we haven't so far succeeded in Western societies in making it other than an administrative-technical operation. This is one dilemma or challenge we face.

A move to more process-oriented, systemic, and comprehensive methodologies is consistent with the emergence of "transfer sciences" (see section 5.4) but goes beyond it. At the present stage of development of the various transfer sciences related to man-nature problems many if not all are badly mired in reductionist concepts and methodology. Perhaps some rapid evolution will occur soon--a prospect that is addressed next.

The model sketched in Figure 10.2 may be sufficiently realistic for purposes of achieving reorganization of the status, substance and process of environmental and resource sciences at universities. It may already have been achieved implicitly in numerous leading insitutions, and there the priority might be consolidation and some trimming of obsolescent components.

In the recent past, many of the university curricula and research programs on renewable resources and environmental topics have developed to bridge the gap between the empiric and pragmatic of Figure 10.2. In many ways these curricula and researches were pre-ecological in essence--many of the professors involved simply showed no real interest in what was happening in the emerging discipline of ecology when broadly defined. Though there was much study on diseases in wildlife, much of that proceeded within taxonomic and physiological contexts. Population dynamics workers usually were more interested in estimating the parameters of simplified pre-cast simulation models than in understanding the population in its ecological setting.

Yet during the past decade or two some initiatives to bridge the gap between the pragmatic and the normative struggled along, e.g. at the Universities of Michigan, Washington, British Columbia, Rhode Island, and at Cornell University. And in the last few years some institutes or programs that appear to be aimed at bridging the gap between the normative and purposeful, with respect to the natural environment and renewable resources, are emerging, e.g. at the University of Washington.

Some schools now appear to have both the conventional SEI System (Figure 10.1) as well as a major complement (Figure 10.2) that has some beginnings of a vertical organization between bridging processes that have strong lateral processes matching their vertical-bridging motivation. Thus with respect to fisheries, the University of Washington's College of Fisheries and its Center for Quantitative Sciences relate predominantly to the empirical-pragmatic bridge. The fisheries resource economists and the Institute for Environmental Studies have important involvements with the pragmatic-normative bridge. The recently organized Institute for Marine Affairs has interest in the normative-purposeful bridge and has apparently used Figures 10.1 and 10.2 (from Jantsch 1972) in its search for a useful niche.

Consistent with Jantsch (1972, 1975) and many others, I recognize the usefulness of heuristic models such as Figure 10.2 for purposes of discussion, debate and consensus building but worry, as they do, lest the models become the framework for action before having been subjected to extensive and intensiveness critical thought.

Still, what is contained in Figure 10.2 is broadly concordant with the contents of previous chapters. The proposal is not so different from the conventional (Figure 10.1) as to frighten many of the conventionalists or to render it unfeasible. In some ways it already reflects what has occurred with the numerous programs, institutes, centers, and new faculties that have emerged in recent decades on environmental and resource matters. Directed activities to mobilize along the lines of Figure 10.2 may now be timely.

10.6 In Conclusion

From a more distant perspective, a standpoint from which the various details of structure and process lose their definition, consider the methodologies and concepts of applied ecology broadly defined as they relate to the emerging planning process.

a) Methodologies: A variety of international organizations, institutes and programs have during the past decade focused on the new scientific methodologies. These include the Organization for Economic Cooperation and Development, the International Council of Scientific Union's Scientific Committee on Problems with the Environment, the International Institute for Applied Systems Analysis, and UNESCO's Man and the Biosphere Program. The United Nations Environment Program has shown strong commitment to support further work. In the United States, the National Science Foundation is supporting a number of studies of interdisciplinary methodologies. Clearly with so much examination already done or being done, any additional new initiative specifically along these lines,

i.e. further rumination, would have to be justified very carefully. It is time to do it and do it well.

b) Concepts: On man-nature problems rather less research and experimental emphasis has been directed toward concepts than to methods. A variety of new international journals has recently sprung up, some of which are quite specifically directed toward concepts. Again a lot has been written recently that has not permeated the ranks of the ecologically-oriented workers and planners.

c) Planning: As already indicated (see Taylor 1976), the technical-administrative aspects of planning are coming to be well understood in western societies. But how to animate the process so that it is politically useful, from the viewpoint of the citizen, is another matter. Clearly the educational subsystem should be more fully involved, as well as the fully non-governmental organizations such as the press and other media.

What seems to be lacking is an adequate consensus amongst various groups of innovators as to the probable practical implications for science and technology of the broad evolutionary transformation now underway. This need no longer be a dim, dark mystery--with some modest redirection of resources it would not be difficult to make rapid progress in implementing the necessary changes, as I have attempted to demonstrate in these ten chapters.

glossary of acronyms

ACMRR	Advisory Committee on Marine Resources Research
EPA	Environmental Protection Agency
FAO	Food and Agriculture Organization of the United Nations
IBP	International Biological Program
ICNAF	International Commission for Northwest Atlantic Fisheries
IIASA	International Institute of Applied Systems Analysis
MAB	Man and the Biosphere Program
NAS	National Academy of Sciences
NEPA	U.S. National Environmental Policy Act
NERC	National Environment Research Center
NSF	National Science Foundation
OECD	Organization for Economic Cooperation & Development
RMIP	Research Management Improvement Program of NSF
TAC	Total Allowable Catch
UNDP	United Nations Development Program
UNEP	United Nations Environmental Program
ZPG	Zero Population Growth

references

Abler, R., J.S. Adams, and P. Gould. 1971. Spatial organization: the geographers view of the world. London: Prentice Hall, 587 p.

Abrosov, V.N. 1969. Determination of commercial turnover in natural bodies of water. Probl. Ichthyol. 9: 482-489.

ACMRR. 1975. Report of the Eighth Session, Advisory Committee on Marine Resources Research. Rome: FAO, FIR/R171 (En), iv + 20 p.

ACMRR. 1976. Indices for measuring responses of aquatic ecological systems to various human influences. Rome: FAO Fisheries Technical Paper FIR/T151, xii + 66 p.

Alabaster, J.S., J.H.N. Garland, I.C. Hart, and J.F. de L.G. Solbe. 1972. An approach to the problem of pollution and fisheries. Symp. Zool. Soc. Lond. 29: 87-114.

Alderdice, D.F. 1971. Factor combinations. Responses of marine poikilotherms to environmental factors acting in concert. pp. 1659-1722 In O. Kinne (ed.). Marine Ecology. New York: Wiley-Interscience.

Balon, E.K. 1975. Reproductive guilds of fishes: a proposal and definition. J. Fish. Res. Board Can. 32: 821-864.

Baum, G. 1975. Religion and Alienation. New York: Paulist Press, v + 296 p.

Bella, D.A. 1974. Fundamentals of comprehensive environmental planning. Engineering Issues, J. Professional Activities, Proc. ASCE, Vol. 100, No. E11: 17-36.

Bella, D.A., and W.S. Overton. 1972. Environmental planning and ecological possibilities. J. Sanit. Eng., Div. Am. Soc. Civ. Eng. 98: 579-592.

Benington, J., and P. Skelton. 1973. Public participation in decision-making by governments. In Government and Programme Budgeting: Seven Papers with Commentaries. London: The Institute of Public Finance and Accountancy.

Bertalanffy, L. von. 1962. General systems theory – a critical review. General Systems 7: 1-20.

Beverton, R.J.H. 1974. Comment. J. Fish. Res. Board Can. 31: 1308.

Borgstrom, G. 1974. The food-population dilemma. Ambio 3: 106-108.

Brett, J.R. 1971. Energetic responses of salmon to temperature. A study of some thermal relations in the physiology and freshwater ecology of sockeye salmon (*Oncorhynchus nerka*). Am. Zool. 11: 99-113.

Burton, I., R.W. Kates, and A.V.T. Kirkby. 1974. Geography. pp. 100-126 In A.E. Utton and D.H. Henning (ed.). Interdisciplinary Environmental Approaches. Cosa Mesa, Calif.: Educational Media Press, iv + 251 p.

Carlander, K.D. 1969. Handbook of freshwater fishery biology. Vol. 1. Ames, Iowa: Iowa State Univ. Press. 720 p.

Carlander, K.D. 1977. Handbook of freshwater fishery biology. Vol. 2. Ames, Iowa: The Iowa State Univ. Press. vii+ 431 p.

Chant, D.A., and H.A. Regier. 1972. A challenge to the traditional Western view of development. Sci. Forum 5: 3-6.

Chevalier, M., and T. Burns. 1975. A field concept of public management. Ottawa: Environ. Canada. Advanced Concepts Centre. 54 p. (Mimeo).

Curlin, J.W. 1973. Environmental indices. Environmental Policy Div., Congressional Research Service, Library of Congress. Washington: U.S. Govt. Printing Office. 46 p.

Davis, J. 1973. Managing our oceans for mankind. J. Fish. Res. Board Can. 30: 1928-1930.

Dee, N., J. Baker, N. Drobny, and K. Duke. 1973. An environmental evolution system for water resource planning. Water Resources Res. 9: 523-535.

Dickie, L.M. 1970. Introduction to Part Five: Food Abundance and Availability in Relation to Production, pp. 319-323. In J.H. Steel (ed.) Marine Food Chains. Oliver and Boyd, Edinburgh viii + 552 p.

Eipper, A.W. 1964. Growth, mortality rates, and standing crops of trout in New York Farm Ponds. Cornell Univ., Agric. Exp. Sta., Memoir 388, 67 p.

Eisenbud, M., and 14 others. 1974. Planning for environmental indices. Report of the Planning Committee on Environmental Indices to the Environmental Studies Board of the Commission on Natural Resources of the National Research Council. National Academy of Sciences and National Academy of Engineering. x + 47 p.

Elton, C. 1927. Animal Ecology. London: Sidgwick and Jackson. (Republished in 1966 in Methuen's Science Paperbacks series, xvi + 207 p.)

Emery, F.E., and E.L. Trist. 1973. Towards a social ecology. London: Plenum Press. xv + 239 p.

FAO. 1974. Population, food supply and agricultural development. Bucharest World Popul. Conf. Background Pap. E/CONF.60/CBP/25: 27 p.

Finkle, P.Z.R. 1974. Realities of environmental management: the case of marine fisheries. Queen's Q. 81: 240-246.

Forney, J.L. 1971. Development of dominant year classes in a yellow perch population. Trans. Am. Fish. Soc. 100: 739-749.

Forney, J.L. 1976. Year-class formation in the walleye (*Stizostedion vitreum vitreum*) population of Oneida Lake, New York, 1966-73. J. Fish. Res. Board Can. 33: 783-792.

Fox, I.K. 1975. Canadian policy development for management of environmental resources. Vancouver: Univ. British Columbia Westwater Res. Centre. 63 p. (Mimeo).

Francis, G.T. 1976. An overview and summing-up the Rungsted Conference. pp. 48-78 In Environmental Problems and Higher Education. Paris: OECD/CERI, 182 p.

Fry, F.E.J. 1947. Effects of the environment on animal activity. Univ. Toronto Stud. Biol. Ser. 55, Publ. Ontario Fish. Res. Lab. 68, 62 p.

Fry, F.E.J. 1971. The effect of environmental factors on the physiology of fish. pp. 1-98 In W.S. Hoar and E.J. Randall (ed.). Fish Physiology, Vol. VI. New York, Academic Press.

Galbraith, J.K. 1964. Economics and the quality of life. Science, July 1964. As revised in J.K. Galbraith, 1972, Economics, Peace and Laughter. Signet Book E4954, 288 p.

Galbraith, J.K. 1973. Economics and the public purpose. Boston, Mass.: Houghton-Mifflin. 334 p.

Grant, G. 1976. "The computer does not impose on us the ways it should be used." pp. 117-131 In A. Rotstein (ed.), Beyond industrial growth. Toronto: Univ. Toronto Press, xii+ 132 p.

Gillette, R. 1971. Trans-Alaska pipeline: impact study receives bad reviews. Science 171 (March 19): No 3976, 1130-1132.

Gregory, B.P. 1975. Science and discovery in France. J. Roy. Soc. Arts, August 1975, pp. 553-564.

Gross, B.M. 1966. The state of the nation: social systems accounting. London: Tavistock Publ. 166 p.

Gulland, J.A. 1972. Population dynamics of world fisheries. Seattle: Univ. Washington Sea Grant Publ. WSG 72-1, 336 p.

Haig-Brown, R. 1974. The salmon. Ottawa: Environ. Can. Fish. Mar. Serv. 79 79 p.

Hamley, J.M. 1975. Review of gillnet selectivity. J. Fish. Res. Board Can. 32: 1943-1969.

Hardin, G. 1976 MS. Pejorism: the middle way. Paper presented at Boston A.A.A.S. Meeting, 18-24 February 1976.

Harvey, H.H. 1976. Aquatic environmental quality: problems and proposals. J. Fish. Res. Board Can. 33: 2634-2670.

Hayes, F.R. 1973. The chaining of Prometheus: Evolution of a Power Structure for Canadian Science. Univ. Toronto Press, Toronto, Ont. xix + 217 p.

Hetman, F. 1973. Society and the assessment of technology. Paris: OECD, 420 p.

Hirsch, F. 1976. Social Limits to Growth. Cambridge, Mas., Harvard Univ. Press. x + 208 p.

Holt, S.J. 1975. Objectives in conserving the living resources of the sea. pp. 41-46 In Critical Environmental Issues on the Law of the Sea. R.E. Stein (ed.). London: International Inst. for Environment and Development. iii + 57 p.

Inhaber, H. 1974. Environmental quality: outline for a national index for Canada. Science 186: 798-804.

Jantsch, E. 1972. Technological planning and social futures. London: Casell/Associated Business Programmes. xiv + 256 p.

Jantsch, E. 1975. Design for evolution, New York: George Braziller. xxvi + 322 p.

Jenkins, D.W., and 25 others. 1974. The role of ecology in the federal government. Report of the Ad Hoc Committee on Ecological Research. Council on Environmental Quality and Federal Council for Science and Technology. U.S. Govt. Printing Office, Stock No. 038-000-00202. xi + 78 p.

Kesteven, G.L. 1972. Management of the exploitation of fishery resources, p. 229-261. In B.J. Rothschild (ed.) World fisheries policy: Multidisciplinary views. Univ. Washington Press, Seattle, Wash. xix + 272 p.

King, A. 1973. Foreword. pp. 5-9 in Hetman, F. Society and the Assessment of Technology. Paris: OECD. 420 p.

Kuhn, T.S. 1970. The structure of scientific revolutions. 2nd ed. Int. Encycl. Unified Soc. 2: 210 p.

Larkin, P.A. et al. 1970. This land is their land. Sci. Counc. Can. Rep. 9: 41 p.

Larkin, P.A. 1971. Simulation studies of Adams River sockeye salmon (*Oncorhynchus nerka*). J. Fish. Res. Board Can. 28: 1493-1502.

Larkin, P.A., and N.J. Wilimovsky. 1973. Contemporary methods and future trends in fishery management and development. J. Fish. Res. Board Can. 30: 1948-1957.

Laszlo, E. 1972. The systems view of the world. New York: George Braziller. x + 131 p.

Lawrie, A.H, and J.F. Rahrer. 1972. Lake Superior: effects of exploitation and introductions on the salmonid community. J. Fish. Res. Board Can. 29: 765-776.

Lawrie, A.H., and S.R. Kerr (eds.). 1976. Natura naturans: a symposium on the Fry paradigm. J. Fish. Res. Board Can. 33: 296-345.

Leopold, L.B., F.E. Clarke, B.B. Hanshaw, and J.R. Balsley. 1971. A procedure for evaluating environmental impact. Washington: U.S. Geological Survey Circular 645, 13 p.

Loftus, K.H. 1976. Science for Canada's fisheries rehabilitation needs. J. Fish. Res. Board Can. 33: 1822-1857.

Loftus, K.H., and H.A. Regier (eds.). 1972. Proceedings of the Symposium on Salmonid Communities in Oligotrophic Lakes. J. Fish. Res. Board Can. 29: 613-986.

Lucas, C.E. 1973. Scientific advice to fisheries bodies. J. Fish. Res. Board Can. 30: 1958-1964.

Lucas, C.E., and 10 others. 1974. The scientific advisory function in international fishery management and development bodies. Rome: FAO, FIR/R142 Suppl. 1 (En). ix + 14 p.

MacKenzie, W.C. 1973. Reconciling conflicts among different economic interest groups in the management of fisheries. J. Fish. Res. Board Can. 30: 2065-2069.

MacPherson, A.G. 1972. People in transition: The broken mosaic, p. 46-72. In A.G. MacPherson (ed.) The Atlantic Provinces. Studies in Canadian Geography. Univ. Toronto Press, Toronto, Ont.

Marr, J.C. 1973. Management and development of fisheries in the Indian Ocean. J. Fish. Res. Board Can. 30: 2312-2320.

Maruyama, M. 1974. Paradigmatology and its application to cross-disciplinary, cross-professional and cross-cultural communication. Cybernetics 17: 136-156, 237-281.

Maruyama, M. 1975. Paradigmatology and its application to cross-disciplinary, cross-professional and cross-cultural communication. World Anthropology (Proc. 9th Int. Congr. Anthro. Ethnol. Sci.) Mouton.

McCracken, F.D., and R.S.D. Macdonald. 1976. Science for Canada's Atlantic inshore seas fisheries. J. Fish. Res. Board Can. 33: 2097-2139.

McHarg, I. 1969. Design with nature. Garden City, N.Y.: Natural History Press.

McHugh, J.L. 1970. Trends in fishery research. pp. 25-56 In A Century of Fisheries in North America. N.G. Benson (ed.). Am. Fish. Soc. Spec. Publ. No. 7. ix + 330 p.

McHugh, J.L. 1972. Population dynamics and fisheries management. pp. 80-92 In Marine Fishery Resources. R.N. Thompson (ed.). Corvallis, Oregon: Proc. Oregon's 1971 National Discussion Forum, Continuing Educ. Publs.

Medawar, P.B. 1976. The strange case of the spotted mice. The New York Review of Books 23: 6, 8, 10 and 11. (April 15, 1976).

Meyboom, P. 1974. In-house vs. contractual research: the federal Make-or-Buy Policy. Can. Public. Adm. 17: 563-585.

Mitroff, I.I., and L.R. Pondy. 1974. On the organization of inquiry: a comparison of some radically different approaches to policy analyses. Public Administration Review 34: 471-479.

Mitroff, I.I., and M. Turoff. 1973. Technology forecasting and assessment: science and/or mythology. Technological Forecasting and Social Change, 5: No 2, 113-134.

Mumford, L. 1970. The Myth of the Machine, Vol. 2. The Pentagon of Power. Harcourt, Brace, Jovanovich, New York, N.Y. 496 p.

Munn, R.E. (Ed.). 1975. Environmental impact assessment principles and procedures. Scientific Committee on Problems of the Environment of the International Council of Scientific Unions: SCOPE Report 5. 160 p.

Needler, A.W.H. 1973. Chairman's summary of the highlights of the Conference. J. Fish. Res. Board Can. 30: 2508-2511.

Nilsson, N.A. 1967. Interactive segregation between fish species. pp. 295-313 In S.D. Gerking (ed.), The Biological Basis of Freshwater Fish Production. Oxford: Blackwell, xiv + 495 p.

Nordenskiöld, E. 1928. The history of biology. New York: Alfred A. Knopf. xiii + 629 + 15 page index. (Originally published as Biologins HIstoria, in three volumes. 1920-24. Stockholm: Bjorck and Borjesson).

Ostrom, V., and E. Ostrom. 1971. A political theory for institutional analyses. pp. 173-186 In F.A. Butrico, C.J. Touhill, and I.L. Whitman (ed). Resource Management in the Great Lakes Basin. Lexington, Mass.: Heath Lexington Books, xiii + 190 p.

Paulik, G.J. 1972. Fisheries and the quantitative revolution. pp. 219-228. In B.J. Rothschild (ed.). World Fisheries Policy. Seattle, Wash.: Univ. Washington Press. 272 p.

Peters, H. 1976. Tautology in evolution and ecology. Am. Nat. 110(No. 971): 1-12

Pimlott, D.H., C.J. Kerswill, and J.R. Bider. 1971. Scientific activities in fisheries and wildlife resources. Sci. Counc. Can. Spec. Stud. No. 15: 191 p.

Pirsig, R.M. 1974. Zen and the art of motorcycle maintenance. New York: Bantam Books, B8880, 406 p.

Popper, K.R. 1972. Objective Knowledge: An Evolutionary Approach. Oxford: Clarendon Press. x + 380 p.

Pritchard, G.I. 1976. Structured aquaculture development with a Canadian perspective. J. Fish. Res. Board Can. 33: 855-870.

Raiffa, H. 1973. A provisional research strategy. pp. 8-33 In The International Institute for Applied Systems Analysis: background information; provisional research strategy. IASSA DIR/G/I (prov.).

Rapport, D.J. 1971. An optimization model of food selection. Am. Nat. 105: 575-587.

Rapport, D.J., and J.E. Turner. 1975a. Feeding rates and population growth. Ecology 56: 942-949.

Rapport, D.J., and J.E. Turner. 1975b. Predator-prey interactions in natural communities. J. Theor. Biol. 51: 169-180.

Rapport, D.J., and J.E. Turner. 1976. Economic models in ecology.

Regier, H.A. 1972. Community transformation — some lessons from large lakes. In Proc. 50th Year Anniversary Symposium, Univ. Washington College of Fisheries. Seattle: Univ. Wash. Publ. Fish. No. 5, 35-40.

Regier, H.A. 1973. Sequence of exploitation of stocks in multispecies fisheries in the Laurentian Great Lakes. J. Fish. Res. Board Can. 30: 1992-1999.

Regier, H.A. 1974a. Fishery ecology: a perspective for the near future. J. Fish. Res. Board Can. 31: 1303-6.

Regier, H.A. 1974b. Fish ecology and its application. Mitt. Internat. Verein. Limnol. 20: 273-286.

Regier, H.A. 1975. To survey, monitor or appraise fishery resources: some general concepts. EIFAC Tech. Pap. (23) Suppl. 1, Vol. 2: 690-703.

Regier, H.A. 1976a. The dynamics of fish populations, fish communities and aquatic ecosystems: a framework of alternative approaches. pp. 134-155 In J.A. Gulland (ed.), The Population Dynamics of Fishes. London: Wiley.

Regier, H.A. 1976b. Environmental biology of fish: emerging science. Env. Biol. Fish. 1: 1-7.

Regier, H.A. 1976c. A scientific analysis of the assessment function, with examples related to aquatic ecosystems. Argonne, Ill.: Argonne National Laboratories, Proc. Symp. on Biological Significance of Environmental Impact (In press).

Regier, H.A. 1976d. Science for the scattered fisheries of the Canadian interior. J. Fish. Res. Board Can. 33:1213-1232.

Regier, H.A., and H.F. Henderson. 1973. Towards a broad ecological model of fish communities and fisheries. Trans. Am. Fish. Soc. 102: 56-72.

Regier, H.A., and F.D. McCracken. 1975. Science for Canada's shelf-seas fisheries. J. Fish. Res. Board Can. 32:1887-1932.

Regier, H.A., and D.J. Rapport. 1977. Ecology's family coming of age in a changing world. Washington: Council of Environmental Quality. Proc. Symp. on Biological Evolution of Environmental Impact (In press).

Regier, H.A., P.L. Bishop, and D.J. Rapport. 1974. Planned transdisciplinary approaches: renewable resources and the natural environment, particularly fisheries. J. Fish. Res. Board Can. 31: 1683-1703.

Regier, H.A., J. Nautiyal, M. Williams, I. Bayly, I.S. Fraser, and R.D. Vandenberg. 1973. Qualitative and quantitative data. Toronto: Man and Resources/ Ontario Committee. Outlook 73(10): 8-10.

Ricker, W.E. 1975. The Fisheries Research Board of Canada — Seventy-five years of achievement. J. Fish. Res. Board Can. 32: 1465-1490.

Ricker, W.E., and E.P. Weeks (ed.). 1973. FAO Technical Conference on Fishery Management and Development. J. Fish Res. Board Can. 30: 1921-2537.

Rotstein, A. (ed.). 1976. Beyond industrial growth. Toronto: Univ. Toronto Press, xii + 132 p.

Skud, B.C. 1975. Management of the Pacific halibut fishery. J. Fish. Res. Board Can. 30: 2393-2398.

Stewart, R.W., and L.M. Dickie. 1971. Ad Mare: Canada looks to the Sea. Sci. Counci. Can. Spec. Stud. No. 16: 173 p.

Strong, M. 1975. We can cure our ailing world. Toronto Star, Jan. 6, 1975, p. C3.

Taylor, C. 1976. Dilemmas of planning. Canadian Forum 61 (No. 661), p. 32.

Vickers, G. 1973. Foreword. pp. vi-ix In F.E. Emery and E.L. Trist, Towards a Social Ecology. London: Plenum Press. xv + 239 p.

Waddington, C.H. 1975. The evolution of an evolutionist. Ithaca, N.Y.: Cornell Univ. Press. xii + 328 p.

Watt, K.E.F. 1974. The titanic effect. Stamford, Conn.: Sinauer Assoc. xiv + 268 p.

Winkless, N., III, and I. Browning. 1975. Climate and the Affairs of Men. Harper Mag. Pr.

Yorque, R. (ed.). 1975. Ecological and resilience indicators for management. Managing the unknown: methodologies for environmental impact assessment. Progress Rep. PR-1. Vancouver: Univ. Brit. Columb., Instit. Resource Ecology, 85 p.

Zahn, M. 1962. Die Vorzugstemperaturen zweier Cypriniden und eines Cyprinodonten und die Adaptionstypen der Vorzugstemperatur bei Fischen. Zoll. Beitr. (N.S.) 7: 15-25.